Literature in Dialogue with the Natural Sciences

Christoph Strosetzki

Literature in Dialogue with the Natural Sciences

Competing Claims from the Early Modern Period
to the 20th Century

Christoph Strosetzki
Romanisches Seminar
Universität Münster
Münster, Germany

ISBN 978-3-662-71318-1   ISBN 978-3-662-71319-8   (eBook)
https://doi.org/10.1007/978-3-662-71319-8

Translation from the German language edition: "Literatur im Dialog mit den Naturwissenschaften" by Christoph Strosetzki, © Der/die Herausgeber bzw. der/die Autor(en), exklusiv lizenziert an Springer-Verlag GmbH, DE, ein Teil von Springer Nature 2025. Published by Springer Berlin Heidelberg. All Rights Reserved.

This book is a translation of the original German edition "Literatur im Dialog mit den Naturwissenschaften" by Christoph Strosetzki, published by Springer-Verlag GmbH, DE in 2025. The translation was done with the help of an artificial intelligence machine translation tool. A subsequent human revision was done primarily in terms of content, so that the book will read stylistically differently from a conventional translation. Springer Nature works continuously to further the development of tools for the production of books and on the related technologies to support the authors.

Planung/Lektorat: Oliver Schuetze
This Palgrave Macmillan imprint is published by the registered company Springer-Verlag GmbH, DE, part of Springer Nature.
The registered company address is: Heidelberger Platz 3, 14197 Berlin, Germany

If disposing of this product, please recycle the paper.

PROLOGUE

The ancient author Protagoras is attributed with the statement that man is the measure of all things—a statement that could be the motto of the early modern engagement with science, with man as the standard and purpose at the center. Where this is not the case, the anecdote of Thales, who fell into the well while he was studying the stars and looking upwards, is cited as a warning. A maid then mocked him: by wanting to recognize the things in the sky, he missed what concerned him personally. Thus, from an early stage, the study of the natural sciences is viewed with concern from a literary perspective.

Goethe's ballad *The Sorcerer's Apprentice* can also be interpreted as a critique of technological innovations. When the sorcerer's apprentice tries out his master's magic spell alone, he transforms a broom into a servant who carries more and more water for him, relieving him of his work. Unfortunately, his creation takes on a life of its own. The sorcerer's apprentice loses control and can no longer intervene in a process that is becoming autonomous: *The spirits that I summoned / I now cannot rid myself of again.* This not only shows the competence deficit of a dilettante, but also the consequence of a science that has become leaderless, which appears as magic in the parable and goes its own uncontrolled ways. In Goethe, fortunately, the expert comes to put an end to the goings-on. Nevertheless, the narrative is more relevant than ever.

That particularly the natural sciences cannot limit themselves and show their own boundaries was seen as a danger by the Neapolitan Giambattista Vico (1668–1744) in the early modern period. He warned against the scientific striving to go into the infinite and boundless, as this would not correspond to human nature. In addition, he feared the overestimation of mathematical models, especially when they are mistaken for nature itself.[1]

---

[1] Vico 1984.

But where can criteria be found for limiting a boundless thirst for scientific knowledge?

In the early modern period, the sciences, including the natural sciences, were ethically legitimized. The question was asked as to their purpose and the goal to be achieved. Goal-oriented thinking was characteristic. In the universe, man stood at the top as the purpose of creation. His dignity would arise from his intellect, with which he was superior to the animal and plant world. People oriented themselves towards antiquity. Knowledge was passed on in texts and acquired through texts. When the printing press was invented, the book gained such importance that the metaphor of the world as a book became common. In the nature of the world, those who were illiterate and therefore unable to read books could read. Wisdom was more important than knowledge of individual facts. Therefore, curiosity was only appropriate for the recognition of the important. It was believed that predicting the future would impair future happiness. The inner world of the psyche and the spirit was more important than the outer world of physical objects. Historical development seemed less determined by progress than by the cyclical return of the past.

The French mathematician and physicist Blaise Pascal distinguished between a systematic-analytical *esprit de géométrie* and an intuitive *esprit de finesse*. This distinction still shapes the 18th century, when Denis Diderot sees the high degree of abstraction in geometry as a disadvantage, as it moves too far away from everyday reality. Diderot contrasts the geometer with the genius who does not think discursively, but intuitively and based on experiences. Such a mind works without straining. It does not observe, but sees what matters. Without laborious studies, it expands. The encyclopedist Jean-Baptiste d'Alembert adopts the paradigm of geometry. Because the constructs of geometry do not necessarily have a correlate in reality, their level of abstraction is particularly high. For a lexicon, he desires definitions for which geometry is the model. A comparable picture is shown in George-Louis de Buffon's demarcation to Carl von Linne. Linne is accused of proceeding too geometrically and thereby losing sight of reality. Rather, one, Buffon believes, must consider the whole of the individual object.

The inventions of gunpowder, compass, and printing press led to a new perception of natural sciences and technology in relation to humanities. People imagined an original state of humanity and questioned whether historical developments have led to positive or negative outcomes. While Thomas Hobbes considered the political aspect, Bernard Mandeville looked at the economic one. Jean-Jacques Rousseau and Voltaire contemplated the moral consequences of technological progress, with the former rejecting luxury, while the latter advocated for it. Starting from a hypothetical original state, the developments caused by the technical-scientific inventions are thus evaluated differently.

The 19th century marks the triumph of the natural sciences. Philosophy of nature transitions into natural science, where the question is no longer what and why something is, but how it works. Deduction is replaced by induction, and goal-oriented thinking by causality. Nature is no longer observed and described, but explored through experience and experiment. The novelist Honoré de Balzac physiologically compares the human world with the animal world and postulates a *unité de composition*. Émile Zola bases his novels on August Comte's positivism and the inheritance theories of Jean-Baptiste Lamarck and Charles Darwin, thus making his novels follow scientific guidelines. In the 20th century, criticism of the dominance of the natural sciences becomes increasingly vocal, first with Ortega y Gasset, then with the Argentinians Borges and Sábato.

This book is based on an expanded concept of literature that also includes dialogues and treatises. It aims to demonstrate how literature appeared to be prioritized over natural science in the early modern period, while a kind of parity was established in the 18th century. In the 19th century, the natural sciences became predominant, and literature oriented itself towards them. In the 20th century, voices emerged criticizing the priority of the natural sciences. Special attention will be given to the discussion of these developments in France and the Spanish-speaking world, although the situation in Germany and England will also occasionally be considered.

So, not every type of science is contrasted with literature. While the Spanish literary scholar Marcelino Menéndez y Pelayo in his essay collection *Ciencia española* (1878)[2] appreciated the national traditions of Spanish science as a whole, including art, theology, political science, and philosophy, we exclusively contrast natural science with literature. In this, we also differ from the Romance scholar Thomas Klinkert, who in turn considers art with science in general. They diverge, he believes, insofar as science is about true and false, while art is about beautiful and ugly. In his interpretation of literary texts, the guiding question is to what extent references to scientific models of thought, principles, rules, and laws are established and thus a view of scientific knowledge becomes clear. Science is traced from the perspective of literature. The significance of scientific paradigms in fictional texts is shown in that they take the position that theology had in pre-modernity, with Klinkert, like Menéndez y Pelayo, including the humanities and philosophy under the term sciences. He wants to understand fictional speech not as a deviation, but as the basis of communication and not think from the reference to reality, but from the side of communicative action.[3] Our concern, however, is not to appreciate the value of fiction, but to analyze the view of literature on the natural sciences.

Comparable considerations were made in philosophy. After Immanuel Kant had established the basis of the natural sciences in the 18th century, his students, the neo-Kantians, were concerned with the basis of the humanities

---

[2] Menéndez y Pelayo 1953–1954, 432.

[3] Klinkert 2010, 10–14, 37.

in the 19th century. Thus, Wilhelm Windelband called them idiographic, describing the individual, in contrast to the nomothetic natural sciences, which aim at the establishment of laws. For Heinrich Rickert, the humanities deal with values that not only exist but also apply. Wilhelm Dilthey sees worldviews as the starting point for the humanities. These result from life experience and condition and explain perspectives of the same thing. Martin Heidegger criticizes technology because it has generalized the worldview of utilization. He ties in with Dilthey when he makes historicity the central theme of his work *Being and Time* (1927). For Hans-Georg Gadamer, understanding becomes the central category of the humanities again, following Dilthey. In *Truth and Method* (1969), understanding for Gadamer is the original form of existence. Because anyone who wants to understand a text makes a draft of the whole text at the beginning, which results from the expectations of a certain meaning of the text.[4] This draft can be constantly revised by further reading. Gadamer's hermeneutics as a doctrine of understanding shows how understanding a text is shaped by a respective pre-understanding and an expectation horizon and runs in a circular manner. When understanding a handed down text, one's own horizon now encounters another, which is conceived by it as foreign. Thus, a kind of merging horizon arises, whereby the historical horizon is simultaneously conceived and, because mediated with one's own horizon, is abolished. In this process, an effect, an application, can result for one's own circumstances.

The concept of the two cultures, the literary-humanities and the scientific-technical, was taken up by the English physicist C.P. Snow in 1959, and he considered it disastrous that no communication was possible between the two. He attributed a pessimistic, past-oriented attitude to the humanities, while the natural sciences were future-oriented and optimistic. This thesis became known due to strong counter-reactions and earned him the position of a parliamentary undersecretary in the Ministry of Technology. However, he presents the view of a physicist and completely omits the representation of the historical development of the antagonism. In contrast, the following will specifically present this history from the perspective of literature.

On the cover image, Caspar David Friedrich depicts a woman looking at a rising or setting sun. We chose this image because it shows the human being as a microcosm in the face of the macrocosm and suggests different attitudes—from romantic contemplative awe to scientific observation.

At this point, thanks are due to the Fritz Thyssen Foundation for the financial support of the creation of this monograph. Maike Dietz, Fenja Kulemann, and David Wick are thanked for their support in the procurement of materials and review of the manuscript. The consistently good supervision of the project by J.B. Metzler publishing house is owed to Oliver Schütze.

---

[4] Gadamer 1972, 251.

CONTENTS

PART I

PRIORITY OF LITERATURE

CHAPTER 1

Nature

1.1 DEDUCTION

In the early modern period, the view of nature was directed towards the general. It was not the concrete experiences, but the basic building blocks that seemed essential. The question was what nature actually is and not how it works in individual cases. From the most general facts of nature, it was believed that all special and individual aspects could be deductively derived. Since these could be found through mere reflection, quantitative confirmation through experimental arrangements was unnecessary. The ancient tradition provided the essential templates for understanding nature in the early modern period. In Spain, it was particularly Aristotle who set the standards for the second scholasticism of the School of Salamanca in the early modern period.[1] His view of nature will be briefly presented, as it illustrates the method of nature research well.

The basis for physics and all other individual sciences is, according to Aristotle, metaphysics. Its subject matter includes such general concepts as principle, element, nature, essence, time, possession, part, whole, or necessity. It investigates what exists, with its properties. Unlike each of the individual sciences, it does not demarcate any part as its domain, but generally deals with the highest principles.[2] If the most general property of every being is the fact that it exists, then metaphysics deals with being, which is common to all. Individual sciences like physics have a subfield and deal with the particular and its accidents, i.e., its random, changing, and non-necessary properties.[3] The science of nature, i.e., physics, must be distinguished from those

[1] At the universities, Aristotle was the authority. Ramis 2024, 57.

[2] Aristotle 1995a, 61.

[3] Aristotle 1995a, 225.

C. Strosetzki, *Literature in Dialogue with the Natural Sciences*, https://doi.org/10.1007/978-3-662-71319-8_1

disciplines that are directed towards action or production. In the sciences oriented towards action, the movement is not in the object of action, but in the actor. Also where production is concerned, the principle of movement lies in the producer and his art. "The science of the physicist, however, deals with that which has the principle of movement in itself."[4]

Aristotle starts from subjective thinking when he deals with the causes, the first beginnings, and the basic building blocks in his work *Physics*. The starting point is the whole, details are subordinate. Words also initially present something whole, which can then be broken down into its components. As an illustration, he mentions the word "father", which the small child initially uses to address every man and only later distinguishes more precisely. He thus starts from basic concepts such as movement, cause, or matter, which is why the general and not the concrete is in the foreground. By performing a kind of anatomy of the central components of nature and asking what is, individual objects are deductively derived from general premises.

Thus, the five elemental substances earth, water, air, fire, and ether form the basic building blocks from which everything is composed. While the ether fills the celestial space, the other elements are on earth, where they differ from each other through heaviness and lightness, heat and cold, dryness and moisture. Fire is warm and dry, air is warm and moist, water is cold and moist, earth is cold and dry. These elements occur in different mixtures in the individual bodies.

Since for Aristotle man is the measure of all things, he has the highest place in nature and is its ultimate purpose. Because he alone has reason, he makes it the most important characteristic. Aristotle's approach is a kind of inventory of what is in nature. The principle of purposefulness implies a priority of the goal or form, so that higher levels incorporate the properties of the lower ones. While the plant can only grow, the animal can grow and perceive, desire, and move from place to place. Man can do all this and also has the capacity of reason. But in all organic forms, Aristotle finds purposefulness. Among the animals, those with blood are more perfect than those, like crustaceans and insects, that have no blood. In the higher developed animals, the principle of form comes from the male being and the form-receiving and material from the female being.

For Aristotle, nature is the entirety of objects that are associated with matter and move or change. Thus, the emergence is understood as movement from non-existence to existence and the passing away as movement from existence to non-existence. In addition, there is a quantitative movement caused by increase and decrease, secondly a qualitative one, consisting in transformation, and finally a spatial movement through change of location. Such movements take place in space and time. While space is limited, time is unlimited. Without a counting human, there would be no number and no

---

[4] Aristotle 1995a, 232.

time, but only movement with an earlier or later.[5] What is by nature has a drive for change, unlike the products of human activity.

A special kind of change is the becoming-reality of the possible or the bringing-to-an-end of the changeable. Change in general occurs through reshaping in a statue, through addition in growth, through removal when a sculpture is made from a stone, through assembly in the case of a house, or through property change when water becomes ice or when something grows or performs a spatial movement. What performs a spatial movement is either set in motion by itself or by another. The latter happens, for example, by throwing, pushing, inhaling, exhaling, or spitting.[6]

The first cause of a change, without which nothing can be recognized, is the why, the goal. Thus, the goal of walking is the preservation of health. Someone walks so that they feel well. Another cause is that something new arises from something that already exists. If the statue is made out of ore or the bowl out of silver, then these are material causes. The material takes on a different form in the statue than in the bowl. So form or model is added as another cause. And finally, something must initiate the change. Someone has recommended going for a walk. A sculptor works on the statue with tools, or the bowl is shaped by hand. What is the cause for the light and the heavy to move to their respective places? It is natural to them. Water is naturally heavy and goes downwards. But potentially, it is light and goes upwards, namely when it evaporates.[7] Among the existing things, those that are there by nature are to be distinguished from those that are made by man, like a bed or a dress. The action of different causes is shown when a human being arises from a human being, but not a bed from a bed.

Aristotle rejects the opinion of those who attribute the origin of the whole world to chance and assume a primal whirl that has separated and arranged the substances. For from the seed of a tree or a human being, not just anything that happens to come about is produced, but a tree or a human being, which is why nothing in the universe happens as a result of chance. Rather, things arise for a reason, either through planning or through natural disposition. If a horse happens to leave a stable and by doing so avoids a misfortune, then one can speak of chance. But this means nothing other than that the cause lies outside the event and the result cannot be related to a because-of-that. At best, one could speak of a secondary cause, which is subordinate to reason and nature. Natural events always occur in the same way or as a rule. And in processes that have a specific goal, what precedes it is done for its sake. However, just as a spelling mistake is possible when writing, so deformities are possible in nature, which are to be seen as failures of that "because of

[5] Aristotle 1995c, 115.

[6] Aristotle 1995c, 52, 54, 18, 51, 174.

[7] Aristotle 1995c, 203.

something".[8] Thus, natural processes always proceed in the same way, unless something disruptive intervenes.

Such views of Aristotle were adopted and commented upon in the Spanish *Siglo de Oro*, that is, the 16th and 17th centuries, e.g., in Domingo de Soto's *Quaestiones super octo libros Physicorum Aristotelis* (1555), Diego de Zuñiga's *Philosopiae prima pars* (1597), or Benito Pererio's *De communibus omnium rerum naturalium principiis et affectionibus* (1562). When in Francisco Suárez's *Disputationes metafísicas* (1597) a distinction is made between substance and accident, between causal and final cause, and between accidents such as quality, habit, time, and place, it becomes apparent that metaphysics appears as the foundation from which physics must proceed, as Aristotle had already demanded.[9] The second scholasticism does not break with the first scholasticism of the Middle Ages, but in view of humanism, geographical and scientific discoveries, and new political circumstances, it must make a reassessment.[10] Suárez does not limit himself to mere commentary. He also sets clear accents through reorganization[11] and systematization[12] of the Aristotelian text. Therefore, the *Disputationes metafísicas* were the standard work for metaphysics throughout Europe at both Protestant and Catholic universities.[13] According to Suárez, metaphysics as First Philosophy has to search for the first principles of cognition and the first principles of being. Since it does not have a restricted subject area, it is to be designated as a universal science of being in general. The question arises whether Suárez oriented himself towards John Duns Scotus, who had coined the term transcendental science and understood by it what exceeds all categorical modes of being. If one assumes that Suárez built metaphysics as transcendental science based on Duns Scotus, then he paved the way to Christian Wolf and finally to Kant, in whom the transcendentals become a priori conditions of knowledge. Then Suárez would have abolished metaphysics as a doctrine of being and transformed it into onto-logic, which is no longer about being, but about what is conceivable.[14]

Suárez assigns a high value to abstraction. The more things are abstracted from the purely material level, the more clearly and precisely they can be understood.[15] In the hierarchy of sciences, metaphysics is therefore higher

---

[8] Aristotle 1995c, 36–45.

[9] Aristotle 1995a, 124–131.

[10] Ramis 2024, 37–38. For the numerous schools within the second scholasticism cf. Ramis 2024, 62–66.

[11] Mendoza 2022, 30.

[12] Moser 1958, 24

[13] Darge 2019, 22.

[14] Darge 2019, 26–29; cf. also Darge 2014, 50.

[15] "Las cosas tienen tanta mayor perfección en su inteligibilidad, cuanta mayor es su abstracción de la materia, y de modo parecido, el conocimeinto es tanto más claro, cuanto es más inmaterial." Suárez 1960–1966, I, II, 13.

than physics because it moves at a maximum distance from physical matter in its formal considerations. Since the concepts of metaphysics are universal due to their high degree of abstraction, it can support any particular science. Even physics, whose subject matter is material objects, can benefit from it. Therefore, it is metaphysical principles that guide the principles of physics or natural philosophy and make possible classifications of material kinds, in which, for example, a distinction is made between simple bodies, composite inanimate bodies, composite vegetative bodies, those with sensory perceptions, and those with intellect.[16] The metaphysical speculations that the subject makes thus form the foundations from which the world knowledge of physics must also proceed.

Speculations are also the starting point of alchemy. However, no logical deductions are derived from these. Rather, it is the subjective imagination that carries the speculations further. The founder of analytical psychology Carl Gustav Jung (1875–1961), who considered alchemy as "proto-psychology", points out that the researching consciousness of the early modern alchemist wrestled with the problems of the substance, and believed to see shapes and laws in the dark space of the unknown, which, however, did not originate from the substance, but from his own soul. Everything unknown is filled by psychological projection; it is as if the soul's background of the observer is reflected in the dark. What he believes to recognize in the substance, however, are initially his own unconscious conditions.[17] So if one starts from the general thesis that there are similarities in the world, one can deductively derive individual comparisons and references from it. Although similarities do not serve any technical use, they promise insights into the essence of nature. Michel Foucault (1926–1984), who held a chair for the history of thought systems at the Collège de France in Paris from 1970, highlights similarity as a characteristic figure of thought for the early modern period when he claims that in the 16th century it meant to seek meaning by bringing to light what is similar. So one had to compile in all available texts and traditions what was seen, heard, or told. Knowledge is therefore not gained through experience, but through commentary and interpretation.[18]

The doctrine of signatures, according to which every element of nature in its form has a recognizable hint of its hidden properties, is an example. The physician and alchemist Paracelsus (1493–1541), whose original name was Theophrastus Bombast von Hohenheim, developed it, drawing on the neoplatonist Stephanos of Alexandria from the beginning of the 7th century. Thus, the seven planets known since antiquity—Moon, Mercury, Venus, Sun, Mars, Jupiter, and Saturn—are assigned seven metals: the Moon silver, Mercury quicksilver, Venus copper, the Sun gold, Mars iron, Jupiter tin, and

---

[16] Suárez 1960–1966, I, I, 14.

[17] Jung 1944, 314.

[18] Foucault 1974, 60, 72, see also 46–77.

Saturn lead. Pliny had associated the planets with light gradations: Saturn was white, Jupiter bright, Mars fiery, the morning star shiny, the evening star shimmering, Mercury radiant, the Moon mild, and the Sun glowing and radiant. Which hidden properties are further recognizable according to the doctrine of signatures is demonstrated using the example of Saturn. It is the furthest from the Sun and is therefore considered particularly slow, which is why it is often depicted as an old man in human embodiments. As the planet of lead, it appears particularly imperfect and close to chaos compared to the gold of the Sun, which is also symbolized by a dragon. The lead is probably responsible for the fact that Saturn stands for everything dark, deep, and black. The saturnine spirit is associated with melancholy and is a sign of the recognition of secret things and for eloquence. Since the seven days of the week also correspond to the seven planets and Saturn is associated with the Sabbath, in the view of early Christianity he represents the Jewish God, who is overcome by the Sun, the planet of Sunday. It should only be mentioned in passing that the number seven allows further analogies, for example to the *septem artes liberales*, to the seven notes of the musical scale with the resulting harmony of the spheres of the seven planets, to the *artes magicae* of the 15th century divided into seven sub-disciplines, or the seven sages of Greece Plato mentioned in the *Protagoras*.[19] In all the cases mentioned, the analogies are products of subjective thinking, which relates different external bodies or ideas and thus distributes additional attributes.

A list of criteria of how images in analogies can correspond to things is provided by Cosma Rosselli in the ninth chapter of the second part of his work on mnemonics, *Thesaurus artificiosae memoriae* (1579). His list is oriented towards rhetorical figures and is as associative as it is symbolic. Correspondences are possible through similarity. These can be found in the substance (man as a microcosmic image of the macrocosm) and in quantity (the ten fingers of God for the ten commandments). They are possible through metonymy and antonomasia (Atlas for the astronomers or astronomy, the bear for the irascible man, the lion for pride); through irony and contrast (the hollow head for the wise); through ancient symbols (the eagle for Jupiter) or through zodiac signs (the sign of the constellation).[20]

The similarities of things, it can be summarized, are therefore not a property of themselves, but the result of subjective comparison and psychological projection. Subjective ideas are thus assigned to objects so that they can be better recognized in terms of their sense and meaning. On another level, but according to the same model, the metaphysics of the second scholasticism of the *Siglo de Oro* also proceeds when it designs categories from the subject, which it declares to be the foundations of the individual sciences, including

---

[19] Stockinger 2004, 26–28.
[20] Eco 2002, 180–181.

physics. Deduction thus takes place from the most general facts of nature, from the basic building blocks such as the first causes or by attributing similarities, from which conclusions about the nature of things in nature are derived.

1.2   NATURE OF MAN

We also start regarding the nature of man with the Aristotelian tradition so important in the early modern period. If the nature of man, like with Aristotle, is defined by his goal as *eudaimonia* (felicity), then this is the basis from which further things can be deductively derived. Secondary is what material components he consists of or what external causes act on him. Since the goal, the felicity, can only be achieved in the state, Aristotle concludes that man, from his goal, i.e., from his nature, is a social being.[21] Furthermore, felicity can only be achieved through virtue, which is defined as the right middle between extremes. Therefore, ethics is the means to achieve the goal that man is naturally inclined to, and therefore an important component in his characterization. Everything that happens for a goal is influenced by it. The goal, as a *causa finalis*, has a causality. The final cause preexists as a thought in the consciousness of man and is at the beginning of the consideration of whether something should be done; only then do the means come into play. The will can only be influenced by what appears valuable or pleasant, i.e., good. The good is defined by Aristotle as that which everything strives for.

The nature of man should therefore not be considered as an object, but as a subject defined by the goal of felicity. Since he can only achieve this goal in society, sociability is a means to achieve this purpose. Furthermore, the goal is characterized by virtue and the good. If the subject is at the center, then its self-understanding is more important than anything else. These are, in summary, the most important theses of Aristotle. It would be easy to demonstrate to what extent they can be found in the representatives of the Spanish School of Salamanca, which belongs to the second scholasticism. More challenging would be the attempt to demonstrate basic structures of the Aristotelian theses in an author like the French essayist Michel de Montaigne (1533–1592), who is usually assigned to the school of skeptics.[22] Montaigne gives extensive information in his *Essais* about what he thinks is relevant for his readers. In doing so, he rejects the attitude of the grammarian as well as that of the humanistic antiquarian, as neither is concerned with the subjective self-understanding that he wants to put at the center. Montaigne considers man as an acting subject, whose literary education should guide to right action.

In contrast to the grammarian, Montaigne does not want to communicate irrelevant details in his writings, but himself in his self-understanding. With

---

[21] Aristotle 1995c, 4.

[22] See here Chap. "Doubts about Knowledge".

Dionysius, he can only make fun of self-forgetting grammarians, whose only concern is the adversities of Odysseus, while they forget their own. The students of the grammarians pass on their knowledge only to draw attention to themselves, without appropriating it. He has contempt for the one who leaves his study after midnight and in the books has not sought how he could become better, more content, and wiser, but only a poetic meter of Plautus and the correct spelling of a Latin word. Montaigne is puzzled by the idea of filling 6,000 books only with grammatical content. He considers it misguided not to let history be written by those who have insight into historical processes, but by grammarians who pride themselves on good style, as if it were about learning grammar. Montaigne therefore cannot appreciate the linguistic and formal efforts of the grammarians because they do not seem to serve self-understanding or text understanding.[23]

No less than the grammarians, the humanists forget that the ultimate goal is felicity. Despite his own admiration for antiquity, Montaigne criticizes the humanistic way of dealing with antiquity. It contradicts his conception of understanding because it only considers the ancient tradition superficially and does not relate it to the observer. The difference between Montaigne and the humanists is evident, for example, in the contrasting evaluation of Cicero. Montaigne is interested in Cicero's moral philosophy, not his style. Because Montaigne wants to become good and wise, not eloquent. Therefore, he considers Cicero's grammatical finesse useful only in school use. Montaigne criticizes that education which is more oriented towards the Greek and Latin language than towards the acquisition of content. Virtues are thereby only linguistically received and not understood in order to act in accordance with them. So it is not about declining "virtue", but about loving it: "Nous sçavons decliner vertu, si nous ne sçavons l'aymer."[24]

The same applies where writers report on great deeds. Here too, it is not the literary achievement that matters, but the moral appeal to emulate such deeds. As already in relation to the grammarian, Montaigne emphasizes here, too, that the transmission of great deeds should not be left to outsiders, but to those who have committed them. Because they understand them best, and the reader understands them best from their pen, regardless of their style. Montaigne therefore likes to read Caesar, who writes about his campaigns from his own experience. Because, what should one think of a doctor who writes about warfare or a student who discusses the plans of princes? Here too, as with Aristotle, Montaigne emphasizes correct action and the imitation of deeds. In general, the goal of one's own felicity should not be forgotten. Therefore, the humanists who live only in the world of books but neglect their own situation are also to be criticized. They are familiar with,

---

[23] "Les autheurs se communiquent au peuple par quelque marque particuliere et estranger; moy, le premier, par mon estre universel, comme Michel de Montaigne, non comme grammairien." Montaigne 1962, 782; see also 137, 236, 923, 397.

[24] Montaigne 1962, 644; see also 393.

for example, the writings of Galen. But they are incapable of applying them to the concrete patient. Those who deal with books professionally are often vain and weak. As an exception, he mentions Adrianus Turnebus, who had no pedantic streak, although he had dealt with nothing but books. But most humanists understand neither themselves nor the books they read.[25]

The presence of one's own subject is also a criterion when Montaigne distinguishes different types of readers. Thus, the representatives of legal scholarship are contrasted with the simple people. While the former absolutize their scholarship, the latter are not able to understand a text. Montaigne names as a third, rarer group, which he wants to address with his *Essais*, those for whom literary education is an addition, an enrichment of the fully developed personality. It should start from one's own self-understanding and be linked to the knowledge that serves moral behavior and its goals. So anyone who wants to deal with the *artes liberales* should first choose the art that makes him free.[26] In general, in literary studies, external determination and understanding of others should be subordinate to self-understanding. Because: "Mon mestier et mon art, c'est vivre."[27] If for Montaigne the right life becomes the main occupation and goal, then the context that should enable this also becomes important for him.

The literary education, with which Montaigne wants to deal differently than a grammarian or a humanist, was accessible in books. Books are where Montaigne wants to encounter foreign opinions. In book form, he makes himself understood. The question arises whether Montaigne's understanding theory is shaped by his engagement with the medium of the book, or whether the understanding of books is derived from understanding in oral conversation. In reference to Horace[28], Montaigne distinguishes only entertaining books from another type, which intersperses entertaining elements with those through which one can learn to moderate one's own moods and optimize one's own situation by giving useful advice. He repeatedly emphasizes the dual meaning that books have for him: In them, he seeks carefree entertainment or, more importantly, advice and self-knowledge. Books, without which he does not undertake any journey, are for him entertainment and at the same time life help. Books, with which one makes an effort, should not merely entertain, but promote self-understanding through the understanding of foreign opinions and serve one's own life.[29]

---

[25] "Le plus souvent ils ne s'entendent, ny autruy." Montaigne 1962, 138; see also 398, 138, 643.

[26] "Entre les arts liberaux, commençons par l'art qui nous faict libres" Montaigne 1962, 158.

[27] Montaigne 1962, 359.

[28] Cf. Screech 1962, 578–579. Regarding Montaigne's positive attitude towards Horace as a "teacher of correct living, of 'recte vivere'", cf. Buck 1981a, 186.

[29] "Nostre grand et glorieux chef-d'oeuvre, c'est vivre à propos." Montaigne 1962, 1088; cf. also 389, 388, 806, 1088.

This goal is missed by the pedants when they draw knowledge from books and only let it reach their tongue to then spit it out. Their knowledge remains in the misunderstood area of memory and is not appropriated by reason. They live only in the memory of their read books and eventually become so dependent that they can do nothing without a book reference in their book-induced complacency and answer every question with reference to a book.[30]

Aristotle defined man as a social being. What significance do dealing with books and dealing with people have for Montaigne? Montaigne puts dealing with books in a very concrete context. He values the diary that wants to record the most important personal events for a circle of acquaintances and family for a few decades. He too claims to want to write only for a small circle and a few decades. Since his book provides information about him more quickly and thoroughly than personal contact could, he wishes to connect with the person who likes his mood descriptions. Here, dealing with the book leads to dealing with the person, the situation of reading to the situation of oral communication. Both are closely related in Montaigne, as the same understanding patterns apply to both. Consequently, proofs in books by the publishers Vascosan or Plantin are no more compelling than what one sees in our village. Montaigne, who knows that one writes as one speaks, does not cultivate a humanistic book cult. He has no understanding for people who only believe what is written in books. For him, everyday communication is not fundamentally distinguished from dealing with books: "Je parle au papier comme je parle au premier que je rencontre."[31]

Society, sociability, and communication have for Montaigne like for Aristotle a share in the achievement of happiness and virtue. Since society and friendship are central values for Montaigne, he seeks communication with others. He can only affirm withdrawal into isolation in order to conduct self-conversations. Oral interaction has stronger effects than dealing with books. One learns not from books, but in conversation. He values oral communication more highly than written. Excessive reading is harmful not least because it leads to a loss of the ability for polite conversation. However, one is on the safer side with books, Montaigne admits, as the value of dealing with people is random and depends on the respective partner.[32]

Depending on the participants, a conversation can be beneficial or harmful. The conversation with strong and disciplined partners is said to be instructive and beneficial. In the other, one's own mistakes and the foreign mistakes in one's own self can be corrected.[33] Here, Montaigne's doctrine

---

[30] Montaigne 1962, 135, 136, 151, 905.

[31] Montaigne 1962, 767, cf. also 960, 959, 1059.

[32] Montaigne 1962, 801, 900, 34, 41, 163, 805.

[33] "Reprendre en autruy mes propres fautes ne me semble non plus incompatible que de reprendre, comme je fay souvent, celles d'autruy en moy." Montaigne 1962, 146.

of understanding actions and norms of action becomes a doctrine of appropriating norms of action. Even though Montaigne does not seek to lecture, his confrontation with foreign and personal behavioral norms inevitably leads to moral confrontation and correction of one's own self, as well as that of the reader of the *Essais*. Because an external storage of norms in memory is rejected and appropriation through understanding is recommended.

Good action is the goal for Montaigne. Therefore, he discusses different actions and motives. He thus lets the description of examples be accompanied by useful advice. And Plutarch's examples from the lives of ancient heroes are only meaningful if they aim to show the way to virtue. Moral advice is also given by his favorite author Seneca. The narration of an event seems pointless to Montaigne if the reader could not apply it as a lesson to himself. He advises his readers to keep ancient authors like Cato, Phocion, and Aristeides as controllers of their own motivations and goals. They are capable of showing the right way and offer true philosophy.[34] In the representation of historical customs and events, Montaigne—similar to a theologian or philosopher—is less concerned with the reproduction of historical truth, but primarily with ethical instruction.

The subject is at the center of interest for Montaigne. Its goals such as happiness, virtue, and the good are what matter. Since these can only be achieved in a society, interacting with others and with books—another kind of dealing with others—comes into focus. It is important not to lose sight of the central goal, as the grammarians and humanists do. Therefore, conversations serve Montaigne not only for entertainment but also for instruction, which is not to be processed by memory, but by reason. Montaigne does not place communication with books above oral communication, he wants to write as he speaks and addresses his book to a small circle of friends. The wrong handling of books can cause harm. The right handling of books and conversation partners deals with the foreign by appropriating it or, in the case of a mistake, a vice, rejecting it. In this case, a moral correction takes place, so that narratives or examples are applied to one's own life.

1.3   Nature as a Book

When observing the world, the sensually perceptible and positively given appears less important than the thought it triggers. The idea that the world can be read like a book appears in the Latin Middle Ages, where preachers are advised to use not only the Bible but also the book of nature as a source of material. Corresponding ideas can be found in Hugh of Saint Victor, Johannes of Salisbury, and Bonaventura. Since the metaphor of the world as a book has been preserved from the Latin Middle Ages to modern times, the Romance scholar Ernst Robert Curtius describes it as a constantly recurring topos. The

---

[34] Montaigne 1962, 242; see also 158, 105, 216, 357, 242, 104.

alchemist Paracelsus, for example, contrasts the printed books with the more perfect book that God himself has given, dictated, and set. The book of the doctor should be the sick and the firmament another book of medicine, from which one should compile a corresponding sentence or substance. And the whole earth is a library that one may travel through as a wanderer, turning the leaves of the books with one·s feet. In France, Montaigne refers to the great and wide world, which he recommends to his student as a textbook. Descartes is also to be mentioned, where at the end of the first part of his *Discours de la Méthode*, the student, after studying literature, only wants to deal with what he finds in himself or *dans le grand livre du monde*. And Diderot and Rousseau recommend reading in the *livre du monde*.[35]

The Münster philosopher Hans Blumenberg ties in with Curtius' topos in *The Readability of the World* (1981), but focuses more on the specific historical meaning of the respective occurrence of the metaphor. First, he asks himself how it was possible that a book associated with semi-darkness, shortsightedness, and dust became a metaphor for its opposite, the world or nature. He gives two reasons for this: Firstly, the competition between the world as creation and the Bible as a creation and redemption report, and secondly, the ability of the book, like the world as a whole, to create totality. In this context, the metaphor of the world as a book not only becomes, as Blumenberg believes, a polemic against the book as the epitome of sterile scholarship, but rather an indication of the revaluation that the book as a means of communication has experienced thanks to printing. Therefore, understanding nature as a book does not mean devaluing the book in favor of the world, but rather attributing such a societal rank to the book that its metaphorical use in connection with the world appears promising.[36]

For the *Siglo de Oro*, the mystic and Dominican Luis de Granada can be cited, who in his *Símbolo de la fé* (1583) recommends reading the wisdom and greatness of the author from the great book of the beautiful and perfect creatures of the world.[37] Especially the illiterate, who cannot read printed books, find an adequate alternative in this way. Quoting Seneca, Luis de Granada recommends exploring the secrets behind visible things through contemplation. By nature, man is interested in secrets. If many traverse the most remote and impassable regions of the world by traveling, they are guided by an interest in hidden things.[38] It is not the superficially visible, but the hidden message behind it that matters. The writer Antonio de Torquemada deals with the concept of nature in his *Jardín de flores curiosas*

---

[35] Curtius 1973, 323–329.

[36] Blumenberg 1981, 17, 64.

[37] "vemos cómo por el conocimiento de las criaturas nuestro entendimiento se levanta al conocimiento del Creador, así como por el conocimiento de los efectos venimos en conocimiento de las causas de donde proceden. Pues como este mundo visible sea efecto y obra de las manos de Dios, él nos da conocimiento de su hacedor." Luis de Granada 1948, 14.

[38] Luis de Granada 1948, 13.

(1570). He discovers in all beings of creation the will and understanding of the creator. To reach this source, he calls for the contemplation of the things of the world. The orderly and harmonious course of the stars, the effects of the sun, moon, and other planets, like the polar forces, testify to a clear conception.[39] Numerous are the elements of the world, from the clouds to humans and their shape to fruits and flowers, which can be seen as divine messages.

The Augustinian monk Luis Alarcón modifies the metaphor of the world as a book in his *Camino del cielo* (1547), by distinguishing two types of phenomena in the world from an Augustinian perspective. The ones he calls "books of God" are the works of God, the others, which he refers to as books of the devil, can be recognized as works and instruments of the devil. If both phenomena also include the corresponding printed books, the distinction serves as a basis for censorship and inquisition. Then, indeed, the good books are distinguished from the bad books. The first group of bad books consists of superfluous books, which are irrelevant for salvation, the second contains immoral texts, which, like the *Amadís* or the *Celestina*, lead to imitation, the third group contains profane false reports and lies with which one wastes one's time, and the fourth is particularly to be condemned because it lies in the religious field and spreads heresies and false doctrines.[40]

Juan de Pineda in turn modifies the metaphor by supplementing the book metaphor with a letter metaphor. The world becomes a primer in which the alphabet of creatures is inscribed. Each individual human serves as a living letter. If one considers this world of creatures, one moves from the letters to the meanings of words, then to sentences, and finally to the understanding of the whole. While the letters are perceived by the eyes, the unfolding of meaning and word meanings becomes the task of higher mental abilities, leading from the individual to the general. The most general thing that can be gleaned from the letter-like elements of the world is the knowledge of their creator. The letter metaphor thus becomes a metaphor of knowledge, with the process of reading corresponding to that of recognition.[41]

So if nature serves as a book, it is not the printer's ink that is interesting, but the meaning of the letters and sentences. Through the perceived, one may be in search of the mysterious that lies behind it. The world as a book serves on the one hand as a collection of material for the speaker. For the illiterate, it is the alternative to the Bible. The well-ordered nature of natural phenomena reveals to the contemplative observer the hidden secrets behind them, of which the secret of the world's creator is the greatest. The evil in the world is to be attributed to the devil and avoided, regardless of whether it manifests itself in visible objects or in printed books. Phenomena

[39] Torquemada 1982, 105.

[40] Strosetzki 1987, 203–209.

[41] Pineda 1963, Vol. 2, 443.

of nature thus become indicators of the messages and teachings they point to. As such and in their material appearance in themselves, they are uninteresting. One does not want to experiment with them, but encounters them contemplatively.

1.4   Natural Law

Since nature is rationally ordered, it was thought that rational laws could also be derived from it. However, the focus was not on mathematically formulated sentences of theoretical physics, but on ethical and legal laws. Since, according to Christian conception, creator and creation possess the same reason, the morally good action of the individual is oriented towards the reason of creation. When actions are viewed from the goal, both morality and law are equally involved. In antiquity, starting from the substance, i.e., the duration as opposed to the changing, a comparable result was reached. With Aristotle, the nature of a thing is known to be its goal and purpose: "For the nature that each thing has at the completion of its creation, we call the nature of the thing in question, whether it be a man or a horse or a house or whatever else."[42] From the nature of a thing or from its definition, it can therefore be deduced how it should be. The essence of a thing is in its immanent form, which brings it from possibility to reality. If man is a *zoon politikon*, then good constitutions are those that enable his development in a community. It should not be forgotten that the degenerate constitutions have lost sight of the middle. The golden mean between too much and too little is a basic principle derived from the proportions of nature, which applies equally to politics and ethics. Courage appears as the middle between recklessness and cowardice, justice as the middle between doing wrong and suffering wrong, while injustice is too much of the advantageous and too little of the disadvantageous.[43]

Aristotle associates law with morality and deals with justice, law, and equity as virtues in the fifth book of the *Nicomachean Ethics*. If the basic commandment for all virtues is the *mesotes*, the golden mean between too much and too little, then injustice is the violation of a proportion, so that, for example, with one's own person, too much of the advantageous and too little of the disadvantageous is injustice.[44] By prohibiting actions, the law generally promotes virtue, as when it prohibits flight in battle, courage is promoted. The Aristotelian concept of law is therefore not detached from the ethical context. Aristotle distinguishes between dianoetic virtues such as prudence and wisdom and ethical virtues, which consist in maintaining the right middle of extremes. Thus, courage is the middle between the non-values of cowardice and recklessness, and generosity is the middle between the non-values

---

[42] Aristotle 1995d, 4.

[43] Aristotle 1995b, 115.

[44] Aristotle 1995a, 115.

of stinginess and wastefulness. The four main virtues, in which all others are contained, have been called cardinal virtues since Plato: wisdom, courage, temperance (moderation), and justice.

The Spanish theorists of the School of Salamanca established a philosophical tradition, which is complemented by a legal one. The latter is based on the representations of Roman law, in which natural law, starting from the *Corpus iuris civilis*, forms a considerable part. The School of Salamanca includes Francisco de Vitoria, Domingo de Soto, Melchior Cano, Alfonso de Castro, Francisco Suárez, Martín de Azpilcueta, Diego de Covarrubias y Leyva, Fernando Vázquez de Menchaca, and Luis de Molina. They all continue to hold a special place in the history of natural law doctrines.[45] Their contribution to the development of the doctrine of fundamental rights is undisputed.[46] Particularly evident is their influence on the much-cited German jurist Hugo Grotius.[47] After all, Grotius had learned about the writings of the Spanish second scholasticism through his teacher in Leuven, Leonardus Lessius.[48] Grotius, in turn, often refers to the "divine natural law" of biblical revelation.[49] His work *De jure belli ac pacis libri tres* (1625) is often mistakenly seen as the beginning of the modern discussion of natural law, which is continued by the legal historian Samuel Pufendorf's *De officio hominis et civis libri duo* (1673).[50] Grotius sees the Christian countries driven by a rage that respects neither divine nor human law. International law, which applies to societies considered as individuals, is also considered natural law.[51]

For Aristotle, the law in the polis is partly natural law, partly statutory law. The former is valid everywhere and is independent of consent.[52] Even if people do not have their own language and have not made an agreement with each other, they have deeply rooted norms, such as those of freedom or the duty to bury deceased relatives.[53] Natural law has a higher rank than the laws passed in the polis for Aristotle. The latter are right or wrong, depending on whether they correspond to natural law.[54] In the Christian Middle Ages, according to Thomas Aquinas, the *lex aeterna* is revealed in two ways: from the conscience of man "per modum cognitionis" or "per modum actionis et passionis" through the "inclinationes naturales", which include self-preservation, reproduction as species preservation, and finally education and

[45] Seelmann 1997.

[46] Köck 1987; see also Reibstein 1958, 333.

[47] Recknagel 2010.

[48] Scattola 1999, 208.

[49] Ertz 2014.

[50] Scattola 1999, 206–207.

[51] Scattola 1999, 210.

[52] Aristotle 1995b, 10.

[53] Aristotle 1999, 63.

[54] Brüllmann 2011; see also Gärtner 2012.

community formation. Positive law is only justified if it is moral, serves the common good, is drafted by the competent authority, and is adapted to the circumstances and mentalities.[55] Even with Thomas, questions arise about the basic content of action norms that are generally valid under all historical circumstances and form a control instance of positive laws.[56] The Spaniard Francisco Suárez ties in with the doctrine of the importance of conscience[57], which enables an insight into the *lex naturalis*. While the *lex naturalis* is unchangeable, the way and manner in which man perceives natural law is subject to change.

When hierarchies of laws are established, the subordinate level is to be deductively derived from the superior one. Thus, one could derive a *ius naturale* from the *ius divinum*. The former, in turn, is supposed to be the standard for the laws of positive law, which emerges from the *ius naturale*. The origins of this derivative relationship lie in antiquity. The Stoics assumed a divine world reason and distinguished between the eternal world law (*lex aeterna*), the natural law (*lex naturalis*), and the man-made law (*lex humana*). If the world law dominates everything, one has to submit to it. If one does not do this, one will remain as unsuccessful as the dog that unwillingly sits on its hind legs and is still dragged along.[58] For the Stoic Cicero, nature and reason are on a higher level, which stands above the laws made by man. From the perspective of the *lex naturalis*, one is therefore able to distinguish good from bad human laws. It would therefore be foolish to believe that "everything that is laid down in the regulations and laws of the peoples is just. Even if they are any laws of tyrants."[59] Unjust human ordinances are no more laws for Cicero than agreements of robbers.[60]

This clear derivation becomes shaky in the early modern period because with the power analyst Niccolò Machiavelli, the divine is separated from the human level and politics is seen as a system independent of ethics and theology. In his view, both teleology and deduction disappear with the elimination of divine and natural laws, while everything is limited to the optimization of means for power expansion.

If the Christian doctrine of creation assumes a common order for creator and creature and thus establishes the derivative relationship between *ius divinum* and *ius naturale*, then the order of the *lex Dei* applies on three levels: in nature, in the laws of the state, and in the moral commands for the individual. If the individual acts morally, then he aligns his individual rationality with

[55] Brieskorn 1990, 106.

[56] Kluxen 2010, 253–262.

[57] The recourse to conscience relates law to morality. An attempt was made to prove the concept of conscience, after its subjectivization and secularization, as a normative boundary concept in positive law: Filmer 2000.

[58] Welzel 1962, 40.

[59] Cicero 2002, 46–47.

[60] Cicero 2002, 48–49.

the reason of creation. From the *ius divinum* a *ius naturale* can be derived, which in turn is decisive for the individual laws of positive law. The latter, as a transient *lex temporalis*, stands in contrast to the *lex Dei* thought of as eternal. The separation of the *lex humana* from the natural and divine law and thus the autonomization of empirical and positive law is also problematized in the legal theory of the Aristotelian Thomas Aquinas. If the positive law were just because it arises from the will of man, then the human will could not be unjust.[61] *Lex humana* should, according to Thomas, be derived as positive law from the two superior levels, *lex aeterna* and *lex naturalis*.

There are individual cases in which civil disobedience against false laws is permitted. It is also conceivable that the entire positive legal system is misguided if it is, for example, in contrast to the human *bonum commune* or to the divine law. For example, finality is not given if a law is not oriented towards the common good, but towards the self-interest of the legislator. The positive laws are also clearly in contrast to the divine good when they lead to godlessness. Here, Thomas categorically justifies the duty of resistance with reference to Paul's sentence that one should obey God more than men.[62] If the end of tyranny is foreseeable and the intensity of oppression bearable, then a revolution can be avoided in the interest of maintaining legal certainty. If the reservations of temporality and intensity are not met, then according to Thomas: If a king abuses his power, he can be "removed from his place or his power can be restricted. [...] And one should not believe that such a people act against loyalty by deposing the tyrant."[63]

In the 16th century, Spain was considered a leading world power and the intellectual center of Europe. The collapse of medieval systems led to the discussion of new orders and conflicts between different norms being particularly intense here. Influenced by the discovery, conquest, and colonization of the New World, the legal philosophical school of Salamanca discussed fundamental questions that still shape international law today. While the 16th century represented an opening for Spain, the 17th century was characterized by attempts to restore old orders. Early modern ideas now clashed with late medieval barriers. However, these conflicts gave rise to a culture of discussion that is unique to the early modern period. Spanish theater has been attributed with casuistic features, as it offers case studies on difficult legal cases involving, for example, honor killings, tyrannicide, popular uprisings, civil and military disobedience, or raison d'état. The genre of the picaresque novel presents protagonists whose criminal actions appear justified. Cervantes' Don Quixote constantly violates existing laws, for example by freeing galley slaves or not paying for others' services, but legitimizes his actions by referring to the higher laws of the knight-errant.

---

[61] Thomas Aquinas 1953, 7–8.

[62] Lippert 2000, 154.

[63] Thomas Aquinas 1971, 24.

The natural law tradition assigns a higher rank to natural law than to law based on enactment. If, for example, it is a natural law to bury the dead, no contrary positive law can change this, as Aristoteles explains in his Rhetoric using the drama *Antigone* by Sophokles as an example: After King Creon decrees that Antigone's renegade brother Polyneices may not be buried according to ancestral custom, but should be left to the birds as prey, he announces as law that those who disobey will be publicly stoned. Antigone, however, feels more bound by natural or divine law than by human law and therefore opposes Creon's law. Comparable examples exist in the Spain of the *Siglo de Oro,* where, for example, in Guillén de Castro's dramas, the resistance of the entire people is directed against tyrannical kings. If the ruler violates Christian morality and natural law, resistance and tyrannicide appear not only justified but essential in Castro's dramatic world.[64] Castro defends the right to resist and to tyrannicide in his works and is critical of the government of absolute monarchy and self-serving rulers.[65] Rebellions and popular uprisings are also numerous in Calderón's theater.[66]

Can general natural laws be applied to specific cases at all? Principles that are particularly general or merely formal are usually consensual as natural laws. A possible solution is offered by the Spanish theorist Francisco Suárez, who belongs to the School of Salamanca, when he assumes different levels of abstraction. He refers to the monk Gratian,[67] who in the middle of the 12th century compiled a systematic collection of ecclesiastical legal sentences under the title "Harmonization of Contradictory Legal Sentences", which became the basis of public law in Europe in the later Middle Ages.[68] Suárez distinguishes between the general and the concrete when he presents three levels of natural reason: first, the general principles of moral action (the virtuous is to be done), second, special principles that reveal themselves through their objectives (justice is to be preserved), and third, conclusions that derive from the principles but are only recognizable through consideration (among the clearer cases is that theft is to be avoided, among the more strongly considered, that usury and lying can never be approved).[69] It is therefore tensions and contradictions that arise from the difference in levels of abstraction.

Just as there is the book of nature alongside the Bible, there are laws given in the Bible, derived from a general reason, that are recorded in the Ten Commandments, alongside the laws of nature.[70] The social duties or

---

[64] Carpotta 1984.

[65] Delgado 1984.

[66] Fox 1986; Lauer 1996, 123–139.

[67] Suárez 2002, 464.

[68] Kaufhold 2013, 97.

[69] Suárez 2002, 460–461.

[70] Ex. 19, 20.

prohibitions of the second table of the Ten Commandments (4–10) appear so universally valid that they were read as a natural law text. Their universality becomes problematic and relativized only in concrete applications, when questions arise such as whether one may kill by law with the death penalty, or how to behave towards parents who abuse their children, or whether one may bear false witness to protect a person. On the other hand, even the Ten Commandments do not appear general enough, so they are elevated to further levels of abstraction in the Bible itself. Thus, the Ten Commandments are repeatedly reduced to the one commandment of love.[71] In response to the question about the content of the law, Jesus answers Luke: love of God and neighbor.[72] Later attempts were also made to generalize. Thus, Benedict of Nursia formulates 74 simple rules for life in and outside the monastery. Drawing on the Decalogue, the concluding sentence is that one should honor all people and do nothing to anyone that one would not want to suffer oneself.[73] A simple plausible formula is also found by Thomas Aquinas, starting from the assumption that the world is an ordered creation and everything strives for the good. From this it follows that everything natural is willed by God. "So this is the first commandment of the law: To do and strive for the good, to avoid the evil. All other commandments of the natural law are based on this commandment."[74] The generalizations of the Ten Commandments were contrasted with the specifications, as found, for example, in the 82-chapter exhortations of Charlemagne for the Christians[75] or in the confessional books of the Spanish *Siglo de Oro*.[76]

In summary, it can be stated that the different levels or degrees of generality can not only lead to difficulties in the relationship between *lex naturalis* and *lex positiva*, but also between the Decalogue and casuistically listed individual cases. However, if one makes the rational laws of nature the starting point, then, as our evidence has shown, one derives legal and ethical laws from them. Already with Aristotle it had been shown that the purpose and goal of a thing arises from its nature. From a definition, it can therefore be derived how a thing should be. If virtues are defined as that which maintains a middle measure, then this also applies to the virtue of justice, with Aristotle as in the School of Salamanca. If every *lex positiva* has to consider the higher-ranking *lex naturalis* and be compatible with it, then human will should not oppose it. But if a *lex positiva* contradicts a *lex naturalis*, resistance against state power is allowed and required, as the theorists and the examples from literature have illustrated.

---

[71] Ad Gal. 5, Apostle 1, Thimo 1.

[72] Lk 10, 26, 27.

[73] Kaufhold 2013, 69–72.

[74] Kaufhold 2013, 107–108. In contrast, William of Ockham asserts that the world has no discernible order. Kaufhold 2013, 123.

[75] Mordek 2012.

[76] Azpilcueta 1570.

CHAPTER 2

Knowledge

2.1 WISDOM AND KNOWLEDGE

Wisdom is associated with knowledge, but knowledge is not always associated with wisdom, for wisdom can provide the criterion by which certain knowledge content is rejected. The wise one is the one who makes the right decisions in his life. Wisdom therefore originates from the subject, while knowledge refers to objects. The wise subject has the criteria for the relevance of objective knowledge. The context can be crucial in the assessment; thus, knowledge in the context of antiquity may be of lesser importance compared to knowledge in the Christian context, or vice versa. The objects of a discipline can make it more or less valuable; it is also possible that a science applies inadequate or dubious methods and is therefore subject to criticism. The methods are subordinate to the goals; thus, astrology was rejected not only because of its methods, but primarily because of its goal, namely to predict the future. One could also believe that a discipline is good in itself, but that the performance of those who practice it is deficient, either because they are not sufficiently trained or because they have a negative attitude towards their work or because vanity, arrogance, and the desire for fame obscure their clear view of the truth. Therefore, they appear to the wise as fools or even madmen, from whom he should distance himself.

In his collection *Antibarbari*, the later *Adagia*, which he repeatedly worked on at the ages of twenty, twenty-five, and fifty, the Dutch universal scholar Erasmus of Rotterdam criticizes the monks for distancing themselves as *religiosi* from the world of antiquity. The key to *Antibarbari* is modesty, which is oriented towards the words of Socrates "I only know that I know nothing"; the more educated someone is, the more modest he is. According to Erasmus, Socrates does not represent an opinion, but only refutes other opinions. Cicero's first book of the Academy seems to be the secondary

C. Strosetzki, *Literature in Dialogue with the Natural Sciences,*
https://doi.org/10.1007/978-3-662-71319-8_2

source from which Erasmus derives his knowledge about the Platonic Academy; in it, Cicero attributes the unification of philosophy and ethics to Socrates. While Cicero is skeptical of all knowledge, Erasmus undervalues his own knowledge.[1]

The idea of a republic of scholars, which in Erasmus' *Antibarbari* is called *"res publica literaria"*, becomes a counter-design against those who close themselves off to the *humanis litteris* or *antiquis litteris*. This republic is not only found in the tradition of the Platonic Symposium, but also in the tradition of the Italian Renaissance academies of Florence, Rome, Milan, Venice, Ferrara, and Mantua. In Spain, it is Vives who speaks in *De tradendis disciplinis* of the *vera academia* as a union of learned and good people.[2]

The Spanish humanist Juan Luis Vives had a very clear idea of the representative of the knowledge disciplines. He refers to him with the Latin word *eruditus*, which has been used for educated intellectuals since Cicero.[3] When he sits in his study, surrounded by books, he writes, undisturbed by external influences, in a balance that should not be disturbed by any passion. But the morality of the scholar is also the prerequisite for the moral goal of his studies. For his concept of wisdom has a Socratic character, as knowledge must be useful for one's own life and serve general education. Thus, the wise should not despise the foolish, but help them. The universality of the scholars' knowledge is broad and should include theology, philosophy, jurisprudence, philology, pedagogy, and psychology. But knowledge always has a moral basis and must be useful for the individual and for society. The moral purpose is then endangered when scholars act not for the benefit of their fellow men, but for their own interests and for the sake of money. Money and scholarship are incompatible for Vives.[4]

The connection between the stoic calm of the scholar and his moral mission to educate fools is attempted to be explained by Vives in chapter nine of his fourth book *De concordia y discordia* under the title *De dignitate et officio sapientis*[5], where he initially highlights the dignity and rank of the scholar. He presents the extraordinary image of a deity dwelling among humans when he has freed himself from the many wanderings of the rule of his passions and approaches the summit of wisdom. From the sublime fortress of reason, he looks down on the wanderers who have to assert themselves as if between barren rocks in harsh storms. He distinguishes himself from them, like the seeing from the blind, and the healthy from the sick. He feels like the one who has reached his harbor after the storm and will help those who swim between the waves as much as possible. He will lower himself and lead others

---

[1] Wedel 1982, 22–24.

[2] Schalk 1971a, 18, 22.

[3] Altervogt 1940, 79–81.

[4] Buck 1981b, 15.

[5] Vives 1872–1890, 367–373.

to where he himself has arrived. However, he will be particularly careful not to fall back into the stormy waves of the ocean. Above all, he must keep away from the "public opinion" and confront what the masses approve of with great enthusiasm skeptically. If the wise man is thus attributed the criteria for the relevance of the respective knowledge and the ability to distinguish wisdom from pseudo-knowledge, Vives would find comparable criteria in the transfer of the quantitative method of natural sciences to the majority decision problematic.

Vives presents three aspects: firstly, the high rank of the wise, secondly, his superiority guided by reason over the large crowd of the ignorant, who are driven by the affects, and thirdly, his task to help the latter. His description of the situation of the wise reminds of the Platonic myth of the cave, in which the wise man rises from the underground depths to the sunlight to return and pass on his new knowledge to those who remained in the cave. According to Plato's theory of ideas, all visible objects of the world depend on conceivable ideas, so that, for example, the respective chair on which one is currently sitting owes its shape to the idea of a piece of seating furniture with an elevated seating surface and feet, thus participating in the idea of the chair.

The Platonic assessment of the rational knowledge of the wise is supplemented by Vives through Stoicism, which wants to suppress the disturbing influences of the passions. According to Vives, the wise man is protected by stoic calm, like a shield that gives him peace and stability. He is neither tempted by emotions to act, nor is he brought down from his high peak by rage. He does not get angry when he is pushed, for the higher he stands, the less it is possible to disturb his peace by influences from lower levels. The first step to wisdom is knowledge of one's own self. A self that is composed of body and soul. The latter must be ruled by reason and not by the body. The wise man is thus recognized by the fact that he avoids anger and all passions.

For Vives there are hierarchies within the disciplines; thus, wisdom is unattainable for those who devote themselves to the forbidden arts such as chiromancy, pyromancy, necromancy, and judicial astrology. According to Vives, the profane wisdom of antiquity should not distract from Christian wisdom, which sees the former as ignorant and understands itself as true knowledge, in which all treasures of science and wisdom are hidden.

2.2   KNOWLEDGE THROUGH UNDERSTANDING

While processes in nature are explained, human actions or views are to be understood. How understanding proceeds in detail is illustrated using the *Essais* of the already mentioned Michel de Montaigne, who not only thinks about this on a general level, but also hints at concepts common in hermeneutics such as pre-understanding, broadening of horizons, and the hermeneutic circle. With the moral application of the understood to one's own circumstances, the understanding of others becomes self-understanding.

Montaigne further asks how to understand correctly when dealing with books and what contribution travel can make to self-understanding.

For Montaigne, life presents itself as a non-uniform, irregular, and multi-faceted movement.[6] For Montaigne, there are no matching opinions, neither among different people nor at different times for an individual. No judgment is always and undisputedly valid. Masks, pomp, and ceremonies obscure the view of the essential, the *substance des choses*.[7] These are well-known characteristics of the Baroque era that Montaigne anticipates here. And yet, he does not seem to want to persist in resigned skepticism when he advises not to mistake appearance and concealment for reality. He recommends paying attention to the shell only after the core has been grasped. Multiformity and constant change make it difficult to understand one's own and others' actions and works, which Montaigne discusses in many parts of his work. The following will first present some of his remarks on the general phenomenon of understanding as well as on quoting and commenting as special forms of understanding literary works. Montaigne's handling of books, which he models on his dealings with people, is then intended to show his self-understanding as an author.

In which areas does understanding move? "Les yeux humains ne peuvent apercevoir les choses que par les formes de leur cognoissance."[8] When the eyes see, they can only recognize with the means they already have at their disposal. The mind forms ideas by referring back to forms that are already familiar to it. Sensory impressions are also interpreted differently depending on their previously known contexts. Montaigne cites the example of a snake's saliva, which is interpreted by humans as threatening poison and by the snake as food, which is why it is futile to search for the one substance in the face of these two meanings. Knowledge is thus pre-structured depending on the situation. One's own prejudices also guide the understanding of foreign writings: "On couche volontiers le sens des escris d'autrui a la faveur des opinions qu'on a prejugées en soi."[9] Even the pre-information about the author's context determines the perspective of the reading. For those who work in writing, he pays attention to style and language, for doctors to medical information, and for lawyers, military personnel, theologians, and ambassadors to what belongs to their field.

Own prejudices can also obstruct the recognition of the foreign. Against this, Montaigne recommends travel as the best school and constant incentive to deal with the unknown and new. The foreign and contradicting own habits have a particular appeal when not judged from one's own standpoint,

---

[6] "La vie est un mouvement inegal, irregular and multiform.« Montaigne 1962, 796.

[7] Montaigne 1962, 615.

[8] Montaigne 1962, 516.

[9] Montaigne 1962, 425. "Car chasque science a ses principes presupposez par ou le jugement humain est bridé de toutes parts." ibid., 522; on prejudice see Schalk 1971b; Gadamer 1965, 182.

but by putting oneself in the other's shoes: "Je m'insinue, par imagination, fort bien en leur place."[10] By thus putting oneself in the shoes of others, one can exclude one's pre-understanding and prejudices. The foreign is also more revealing in a moral sense than the similar to one's own: The bad example offers a better moral lesson than the good.

The completely foreign, entirely different from one's own self, can no longer be understood. However, it is advisable not to hastily dismiss something as completely foreign. For through analogy, even animals can be understood. After all, they also have characteristics that are similar to ours and from which conclusions can be drawn by comparison. After all, tribes that revered a dog as their king attributed meaning to its barking and movements. Strangely enough, the more something is removed from understanding, the higher it is valued. This applies to everything that is suspected and guessed from afar, such as the teachings of alchemists and fortune tellers. In unknown territory, fraud is given free rein. Also, anyone who rises above the crowd can no longer be understood: His inequality makes him unsociable. For the same reason, Montaigne considers interaction with a madman impossible. In the search for the new, different from one's own, he therefore advises not to go too far. However, in all understanding of the foreign, one's own should preserve a free space, "une arriere boutique toute nostre [...] si privé que nulle acointance ou communication estrangiere y trouve place."[11] The own should not be given up, but rather modified and reinforced.

Obviously, according to Montaigne, there are different intensities to absorb foreign things. One can imagine them, deal with them, or appropriate them.[12] Understanding is not satisfied with merely acknowledging foreign experiences. Against the background of one's own, they should also be judged and evaluated: The own horizon of experience—as one could explain with Gadamer—should be related to the foreign. Both horizons should merge; the newly gained understanding can now be applied to the modified own horizon in its respective situation: "Ce n'est pas assez de compter les experiences, il les faut poiser et assortir et les faut avoir digerées et alambiquées, pour en tirer les raisons et conclusions qu'elles portent."[13] The melting of the own horizon with the foreign is demanded by Montaigne with urgent metaphors.[14] The foreign, therefore, cannot be understood if it is not related to the own. The merging horizons go so far according to Montaigne that the foreign, once appropriated, is perceived as one's own. Whoever has made Xenophonand Plato his own, his opinions are no longer

---

[10] Montaigne 1962, 225; see also 900.

[11] Montaigne 1962, 235; see also 446, 430, 640, 213, 258, 798.

[12] "Les unes, il les luy [âme] faut seulement presenter, les autres attacher, les autres incorporer.« Montaigne 1962, 986.

[13] Montaigne 1962, 909.

[14] "science, laquelle, pour bien faire, il ne faut pas seulement loger chez soy, il la faut espouser" (177) and "j'aime mieux forger mon ame que la meubler.« Montaigne 1962, 797.

those of the authors he has read, but his own. What Montaigne remembers after reading is that what he no longer knows came from someone else.

The understanding of the action of an individual according to Montaigne cannot occur directly. Just as a doctor should assemble numerous symptoms, circumstances, and considerations as elements to form the overall picture of a disease, so too is each individual action only to be understood against the background of the entirety of the circumstances and the acting person: "Quand on juge d'une action particuliere, il faut considerer plusieurs circonstances et l'homme tout entier qui l'a produicte."[15] Anyone who wants to understand another should carefully seek information about him. Without the idea of an entirety, however, even the collection of elements is pointless: "Il est impossible de renger les pieces, à qui n'a une forme du total en sa teste."[16] As indispensable as the idea of an entirety, on the other hand, is the consideration of each individual element. Only for this reason could Amyot's translations of Plutarch succeed in detail, because Amyot had formed his own overall idea of Plutarch after long and intensive engagement with the text. Thus, understanding foreign works and actions is circular insofar as the whole and the elements are in a conditional relationship.

Not least, it is this circle of understanding that makes access to reality appear problematic in Montaigne. For the foreign is corrected by the familiar, and the familiar by the foreign, by constantly renewing a whole that changes elements, which in turn redefine the whole. In the face of diversity and multiplicity and the constant replacement of appearance by new reality with the help of imagination, certainties fade.

These general explanations about understanding apply equally to Montaigne for understanding foreign actions as well as books, since the book is nothing more than the action of a creative author. Montaigne often identifies himself with his own book. It is not only his creation, he himself is its creation. In general, reading a book also allows acquaintance with the author. From Plutarch's writings, one can deduce the author himself, even his emotional life.[17] Although we can discover Cicero's very private emotional states in his writings, the lives and works of authors should be judged separately. Acquaintance with the author is especially not useful when the work points beyond him and contains elements that he was not even aware of. The reader can be wiser than the author and recognize further meanings.[18] Interpretations can then help in understanding literary works. Real aids are rare. Interpretations that are detrimental to understanding are all

---

[15] Montaigne 1962, 406.

[16] Montaigne 1962, 320.

[17] Montaigne 1962, 694; see also 648. On dealing with books see Buck 1968, 133; Friedrich 1967, 47.

[18] »des sens et des visages plus riches« Montaigne 1962, 126.

too common. Since it is easier to write comments about books than to write books, there are more commentators than authors.

Especially the writings of antiquity have often been commented on to bring them closer to contemporary understanding, as they are interpretable in many ways. However, the task of making religious, legal, or historical texts understandable is no less. The medieval doctrine of the fourfold sense of scripture[19] seems to be still present when Montaigne attributes several meanings to most of Aesop's fables and wants to recognize other, more essential meanings in addition to the mythological ones. For Titus Livius he recognizes a grammatical and a philosophical interpretation, but also others. If Homer—like an oracle—is suitable for various interpretations and representatives of theology, jurisprudence, the military, and philosophy as well as other disciplines can refer to him, then Montaigne seems to have replaced the doctrine of a fourfold sense of scripture with his doctrine of understanding; it structures understanding from the respective prejudice, pre-understanding, and horizon and therefore arrives at various, each different results. Montaigne often illustrates the fundamental openness of a work for subjective understanding with the image of the oracle and the ambiguity of prophetic language, whose authors have avoided a clear meaning so that posterity can attribute the meaning to their words that pleases them.[20]

In addition to understanding profane ancient texts, Montaigne also addresses the understanding of sacred texts. Although divine word can only be heard through the mediation of human speech, theologians should avoid the mistake of writing about divine matters in all too human form. A theological interpretation of the Bible can cause even more controversies than the Bible itself. Thus, Luther sowed more doubt with his explanations than he brought clarity. Montaigne himself distinguishes three levels of Bible understanding: a first one of simple people who believe reverently and obediently, a second one of those who have fallen into error and orient themselves with moderate intelligence on the literal meaning, and a third one of those who are able to fathom the deeper meaning in the Bible.[21]

Understanding in the legal field occupies the jurist Montaigne less in the *Essais*. The meaning of laws seems to be obvious when, given the general similarity of circumstances, a situation can be interpreted and grasped by comparison and experience.[22] He attributes legal and theological disputes to grammatical ambiguities, from which the debates about the correct understanding of the laws arise.

---

[19] Lubac 1959.

[20] Montaigne 1962, 553, 390, 155, 570, 45.

[21] Montaigne 1962, 308, 1046, 299.

[22] Montaigne 1962, 1047; see also 508.

Montaigne also applies his theory of understanding to historical and geo-graphical areas. It is obvious to compare the function he attributes to customs and habits with that of the *préjugé*. Like prejudice and pre-understanding, Montaigne also affirms the collective habits of thinking and acting. "It is through the intermediary of custom that each one is constant in the place where nature has planted him."[23] So customs shape pre-understanding. It would be curious to imagine a people whose actions in housework, marriages, gifts, sales, and purchases are guided by customary rules, without these being known or written down. Even more curious would be a situation without the usual customs, e.g., with young courtiers from whom the usual courtly inter-action with their peers was taken away.[24]

Montaigne wants to demonstrate the thinking habits of peoples in the present and past to his readers, so that they can sharpen their judgment through the idea of the constant change of things. If successful, the result could be described as a broadening of horizons. [25] The historian also strives for a broadening of horizons when he selects from the sheer abundance of material what seems worth knowing to him, when he describes to people the diversity of their living conditions and thinking possibilities in large and small. The fact that he appreciates and accepts every peculiarity in the differences between one people and another does not prevent him from comparing the habits in Mexican Cusco with those in ancient Europe.

Travel is also a means of broadening of horizons for Montaigne, of appro-priating the foreign. He criticizes those who do not travel at all out of fear of the foreign or who shield themselves from everything unknown during the journey. Anyone who is frightened by everything that is contrary to what he is used to seems to be out of his element as soon as he leaves his own vil-lage, and seems to want to protect himself from the infection of unknown air. Montaigne does not only refer understanding to foreign peoples or past times. Even in everyday life, it does not seem easy for him to interpret the actions of others. What some consider a sign of respectability, others see as stupidity. Especially in the political field, actions are interpreted as often as they are different. Given the instability of the individual, it would be wrong to classify each individual action under an assumed constant overall idea. It is better to go from element to element than to hastily assume an entirety.[26] Even in everyday interaction, being and appearance should be kept separate. Appearance, office, and wealth often lend a meaningful appearance to empty words. If the personality is sufficiently complex, Montaigne praises it, as it can adapt to each conversation partner, not only talking to the neighbor about his house or hunting, but also finding a topic for the gardener or the carpenter.

---

[23] Montaigne 1962, 1458; see also 116.

[24] Montaigne 1962, 116; see also 964.

[25] Gadamer 1972, 286; see also Montaigne 1962, 285, 397, 964, 887.

[26] Montaigne 1962, 316; see also 964, 568, 998, 315, 909, 799, 1066.

Conversation partners should have a point of contact, an intersection of their respective horizons, to make their conversation pleasant.

However, Montaigne does not count himself among those who constantly seek flattering confirmation of their own ego in interaction. As with reading texts about foreign people and nations, he also wants to subject his prejudices to critical examination in conversation. Therefore, contradiction is as welcome to him as the judgment of others. He would not fend off every justified or unjustified criticism. Despite all skepticism about the possibility of moral improvement, the criticism of the mistakes of others is simultaneously the criticism of one's own, and the criticism of one's own mistakes helps others in correcting theirs.[27] Already here, moral implications of Montaigne's understanding doctrine are hinted at, which will be discussed in more detail later.

Understanding of others, as extensively analyzed as it is, is subordinate to self-understanding in Montaigne. He outlines the understanding of his own ego with similar hermeneutic terms as the understanding of others. Thus, he also considers here the relationship between elements and the whole and constantly changes his self-understanding, finding the reality of his own ego behind the external appearance.[28]

It has been shown that Montaigne's general hermeneutic remarks can be found applied concretely in various fields. He applies them to the interpretation of ancient texts and the understanding of the Bible; he reflects on the legal and linguistic side of understanding. The interpretation of historically distant actions, the understanding of geographically distant peoples through readings and travels, occupy him as much as the understanding of the people in his surroundings with whom he interacts daily. Montaigne's understanding of the foreign is derived from and guided by his attempt to understand his own ego. He lays down the path of his self-discovery in the *Essais*, which in turn can lead the reader to self-understanding or self-correction.

Commenting and quoting are forms of understanding that Montaigne not only practices, but, as already mentioned, also discusses extensively. He criticizes the comments of philologists because they only replace one word with another, which may be even more unknown, thus doubling the doubts. Instead of clarifying the unclear text, comments only add new texts that in turn require commentary.[29] Instead of progress or an end to the

---

[27] Montaigne 1962, 901, 941; "les reproches que nous faisons les uns aux autres [...] sont ordinèrement contournables vers nous" (907), we "detestons en d'autres les defauts qui sont en nous" (908) and on the other hand: "Publiant et accusant mes imperfections, quelqu'un apprendra de les craindre" (899).

[28] "Je m'estalle entier [...] chaque piece en son siege« Montaigne 1962, 359; "je n'ay rien à dire de moy, entierement, simplement [...] Distinguo est le plus universel membre de ma logique" (319). "je donne à mon ame tantost un visage, tantost un autre [...] Si je parle diversement de moy, c'est que je me regarde diversement" (319); "Ce ne sont mes gestes que j'escris, c'est moy, c'est mon essence" (359).

[29] Montaigne 1962, 1046; "Les glosses augmentent les doubtes et l'ignorance, puis qu'il ne se voit aucun livre, soit humain, soit divin, auquel le monde s'embesongne, duquel l'interpretation face tarir la difficulte? Le centiesme commentaire le renvoye a son suivant" (1044).

need for interpretation, comments only bring about a darkening of insights. Montaigne's own way of reading makes philological comments superfluous. Emerging difficulties cannot spoil his enjoyment of reading. He simply continues reading and tries once or twice more from the then known overall context, making use of the hermeneutic interaction of element and whole, to open up the unclear spot. But if in the opposite case the text initially appears clear, then there is the danger of discovering ever new difficulties if one lingers too long at one spot. To prevent this, Montaigne advises the reader to read so briskly that he does not sink into shallows. If a book becomes an annoyance because of its difficulty, simply put it aside. Not memory, but judgment should benefit from the readings. Since Montaigne unsystematically, aimlessly from time to time, takes one, then another book in hand to leaf through, and interrupts his reading by meditating, the comment of a single dark spot is of no interest to him. Nevertheless, he writes comments himself, but initially for his own needs. For the case that he takes a book he has already read back into his hands, he has adopted the habit of recording the basic ideas and writing them down as marginal notes.[30] Such a basic idea, an "idée generale", can then serve him as prior information, as pre-understanding for a renewed reading.

Further forms of understanding are quotation and takeover. Just as Montaigne does not acknowledge understanding which merely stores the foreign as foreign in memory and does not make it its own, so he also rejects quotations that remain foreign to the quoting author. However, he affirms them when the quoting author has made them his own. Anyone who parrots words and thoughts of other authors or adorns themselves with borrowed plumes and takes over entire passages from foreign books or commissions their own books and has them written by others, meets with Montaigne's disapproval.[31]

Montaigne himself feels rather impaired and interrupted by the presence and memory of foreign books when writing. He advocates letting others have their say in his writings if he shares their opinion or if they express his opinion, and especially if they express his opinion better than he himself. "Car je fay dire aux autres ce que je ne puis si bien dire."[32] The sources of the quotations are not of interest. It is sufficient to make the thoughts one's own. Here, the image of bees that seek flowers everywhere, but then make their own honey from them, applies. The quoting Montaigne has absorbed the opinions and ideas taken from foreign authors to such an extent that he can no longer distinguish them from his own and his critics may rail against him

---

[30] "d'ajouter au bout de chasque livre [...] le jugement que j'en ay retiré en gros, afin que cela me represente au moins l'air et l'idée generale que j'avois conceu de l'autheur en le lisant« Montaigne 1962, 398; see also 389, 1045, 806.

[31] Here applies: "ils cherchent à se presenter par une valeur estrangere." Montaigne 1962, 146, 136, 145, 1033.

[32] Montaigne 1962, 387.

and not against the quoted author. Since the identification with the quoted author simultaneously transfers responsibility for the thought to the quoter, it explains why Montaigne considers the search for the sources of his takeovers unnecessary. Montaigne thus distinguishes two forms of commenting and quoting. He rejects the comments and quotes that remain foreign to understanding. He affirms those that the commentator or quoter makes his own, as they correspond to his theory of understanding.

As has been shown, Montaigne's general remarks on understanding can also be found in the discussion of individual questions. Despite his associative and abruptly structured *Essais*, our compilation in the text and in the history of its creation brought forth a consistent attitude. Montaigne's theory of understanding refers to the self, the other in everyday oral communication, but also to the understanding of literary, legal, religious, and ancient texts, as well as the actions of foreign peoples and historical figures. It addresses the ambiguity and openness of texts, the importance of pre-understanding and prejudice, the interplay between element and totality and thus the circularity of understanding, the merging horizons of the foreign and the own, as well as the appropriation of the foreign and its application to one's own situation. In text comprehension, it grants the interpreter greater insight than the author, on the other hand, it knows the possibility of empathizing with the author in order to understand the text. It sees the reasons for misunderstanding in language and the reasons for incomprehension in the fact that one pays too little attention to it.

Insofar as Montaigne seeks the understanding of the own and the foreign through the book, he understands himself through his book and others through their books. He himself understands himself through the books of others in the same way that others should understand themselves through his book. Such different reception and production relationships are shaped by the same theory of understanding, as Montaigne's remarks on commenting and taking over foreign texts and his evaluation of dealing with books have shown. In this context, Montaigne always rejects the case in which the foreign is not appropriated according to his theory of understanding. Since he seeks his self-understanding through his book in confrontation with other books, he must understand himself as a writer and find this self-understanding in contrast to other types of writers known to him. As writers, neither the grammarian nor the humanist share Montaigne's conception of understanding as appropriation. Understanding their mistakes defines him as a writer.

This understanding and rejection of foreign mistakes is what characterizes Montaigne's search for his own self-understanding. Therefore, his theory of understanding, by demanding confrontation and appropriation or rejection and finally application in one's own situation, often gives the *Essais* a normative, even instructive character. Regardless of this, the theory of understanding is already normative in that it criticizes other forms of knowledge. The same applies to dealing with books in the comparable context of everyday communication with others.

Montaigne is skeptical of those who do not follow his theory of understanding, thus they cannot reach any meaningful knowledge. Hugo Friedrich
is right when he says that Montaigne's skepticism does not paralyze or
obstruct, but rather opens up. However, it does not open up "by refraining
from taking sides for one of the traditional interpretations of the world"[33],
but by the fact that his theory of understanding opens up access to these
interpretations of the world in the first place. The abstention from taking
sides can only be a possible consequence of a relativistic hermeneutics that
would equally validate every truth. With Montaigne, however, hermeneutic
relativism stops and his partisanship begins where it concerns the theory of
understanding itself and its implications. Like Descartes, who, after doubting
everything, found a foundation in the *cogito* on which he could base the rest
of the world, so Montaigne finds in his theory of understanding a basis from
which he can define not only his relationship to himself and others, but also
to the books he writes and reads.

2.3   CURIOSITY AND DISINTEREST

How should one behave in the face of the seemingly infinite amount of
knowledge? Should one curiously absorb everything or should one turn
up one's nose, select only a few things, and meet the rest with disinterest?
Encyclopedic is the approach that includes the entirety of knowledge and
meets any phenomenon with curiosity. In contrast, the approach appears
dogmatic that, based on a religious or philosophical postulate, highlights a
field of knowledge as solely relevant and thus stands in disinterest towards
all other areas, considering them irrelevant. The eclectic philosophy, which
selects the true or most probable from all systems, was rejected by Georg
Wilhelm Friedrich Hegel as syncretism and compared in Wilhelm Traugott
Krug's *Allgemeines Handwörterbuch der philosophischen Wissenschaften* (1832)
to a "patched-up beggar's garment".[34] Justus Lipsius (1547–1606) and
Johannes Vossius (1577–1649) had seen it differently. Lipsius was the first to
use the term "philosophia electiva", and Vossius had written a historical overview of the ancient schools of philosophy in his work *De philosophorum sectis*
(1657). Also Christian Thomasius titled his historical outline of philosophy in
the opening chapter of his *Introductio ad philosophiam aulicam* (1688) with
De philosophorum sectis and comes to the assertion that the most outstanding
philosophers have always been eclectics.[35] In eclectic philosophy, where facts
and topics are selected from history for practical use, individual dogmas lose
their claim.[36] It is therefore history from which the eclecticist takes what he

---

[33] Friedrich 1967, 125.
[34] Dreitzel 1991, 281–343.
[35] Holzhey 1983, 20.
[36] Schmidt-Biggemann 1988, 7–60, 203–222.

recognizes as good. Hermann Lübbe even goes so far as to speak of eclecticism as a central intellectual virtue of the Enlightenment[37] of the late 18th century.

Another variant of encyclopedism is polyhistorism, which has its origins in rhetorical education. As Cicero states in *De oratore*, "nemo poteris esse omni laude cumulatus orator, nisi erit omnium rerum magnarum atque artium scientiam consecutus" (I, 6, 20). The universal knowledge of the speaker is nothing more than a general knowledge that must suffice for the respective listeners and the topics of speech. Here, an encyclopedic tendency is evident, as in Johann Amos Comenius (1592–1670) and his *Buch der Pansophie*, or in Daniel Georg Morhof's (1639–1691) encyclopedically spreading, posthumously published *Polyhistor*. When Gottfried Wilhelm Leibniz finally polemicizes against the Cartesians, the followers of Descartes, under the influence of the Frenchman Daniel Huet in 1673, he accuses them of focusing one-sidedly on rational truths and neglecting factual truths, for the recognition of which the philological-historical disciplines are indispensable. He also emphasizes the value of classical philology for theology.[38] Here, the Cartesians seem to dogmatically oppose the universalism of the philological method. Leibniz himself is universal in the sense that he allows the opposing parties their rights, as he always thinks of something when reading that defends or excuses the writers. He therefore makes it a principle to despise nothing, thus contrasting with four types of dogmatically one-sided scholars: the philologists, the representatives of mathematics and the natural sciences, the scholastics, and the Cartesians.[39] While the philologists arrogantly despise the scientific works of non-philologists as pedantic, the mathematicians and experimental scholars believe only they provide exact and useful contributions, while all others have nothing more than figments of imagination and mere opinions. The scholastics, however, see themselves as guardians of religion and virtue and condemn all others as deviants, but in their strict method, they themselves become pedants. The Cartesians, who arrogantly look down on all scholars who only deal with a specific field, appear just as one-sided, as these in turn look down on the Cartesians. Against these dogmatisms, each excluding the other, Leibniz opposes a universalism that not only recommends the study of various fields of science, but also advises knowledge of all, even opposing directions and party opinions. For it is precisely the overview of different fields of knowledge that leads to the emergence of new truths.

In the following, some dogmatisms are presented in which the decision for a priority devalues or excludes everything else. Since Augustine plays a pioneering role in the religious context, he will be presented in more detail at the beginning. In his early writings *De libero arbitrio* and *De vera religione*,

---

[37] Lübbe 1981, 43–44.

[38] Mahnke 1912, 16–17.

[39] Mahnke 1912, 26–31.

he seeks the true wisdom. First, he reviews the people who consider military service, agriculture, money-making, knowledge of God, or duty fulfillment as the crown of wisdom. Augustine wants to skip the countless sects, each of which prefers its followers to the rest and boasts alone as wise.[40] The argument is now that all people strive for a happy life, but everyone only becomes happy when they see and grasp the highest good. It is like the many objects illuminated by sunlight which depend on only the one thing, the sun. So the rhetorical question arises: Can you deny that the incorruptible is better than the corrupt, the eternal better than the temporal, the inviolable better than the violable?[41] From this, the advice logically follows to be free from the love of changeable things and to consider the highest, i.e., God, as the object of wisdom. The games of the jugglers and magicians arouse curiosity, which wants to expose the deceptions and tricks. Not in this case, but only when it seeks the one wisdom, is curiosity good for Augustine. The right curiosity strives for the knowledge of eternal, constant things.[42] However, he considers people to be miserable, "who become tired of the known and only rejoice when something new comes, who prefer to learn rather than know, while knowledge is the purpose of learning. [...] Therefore, those who desire the goals themselves, first lack curiosity, knowing that only the inner knowledge gives certainty."[43] The consequence of this priority is disinterest in the world. Augustine had picked up the idea of the priority of the self over the world to accuse natural scientists of counting stars and grains of sand on the beach, measuring celestial regions and calculating star orbits, and forgetting themselves in the process.[44] Augustine admonishes that one should hurry as quickly as possible to where God calls us through his wisdom. One should not love the world, "for all that is in the world is lust of the flesh, lust of the eyes, and pride of life," Augustine warns, citing the Bible.[45] When he urges us not to be distracted by visible spectacles and   theatrical performances in order not to deviate from the truth and to befriend shadows, he seems to be alluding to Plato's allegory of the cave. In the interest of truth, therefore, disinterest in the visible world is required.

[40] Augustine 1962, 157.

[41] "Quid? Incorruptum melius esse corrupto, aeternum temporali, inviolabile violabili poteris negare?" Augustine 1962, 164–165.

[42] Augustine 1962, 535.

[43] "Quibus cognita vilescunt et novitatibus gaudent, libentius discunt quam norunt, cum cognitio sit finis discendi, [...] Quare qui fines ipsos desiderant, prius curiositate carent, cognoscentes eam esse certam cognitionem quae intus est," Augustine 1962, 537.

[44] Augustine 2000, 118–119.

[45] "Quoniam omnia quae in mundo sunt, concupiscentia carnis et concupiscentia oculorum est et ambitio saeculi" Augustine 1962, 543; 1. John 2, 15–16.

Indeed, the second scholasticism, shaped by Aristotle and Thomas Aquinas, dominates at the Spanish universities of the 16th century.[46] In contrast, authors of mysticism and spiritual literature often refer to the Platonism of Augustine. As an example, the work *Libro de la vanidad del mundo* (1562) by the Franciscan Fray Diego de Estella (1524–1578) is cited. Like Augustine, he is convinced of the priority of the eternal, in contrast to which the world appears vain. He demonstrates how much disinterest in worldly knowledge is required by looking at lawyers. After all, legal scholars were the fiercest opponents and persecutors who opposed Christ. This shows that a malicious lawyer can cause more damage than a hundred ignorant people.[47] What applies to worldly knowledge also applies to worldly books. It is better not to be interested in them and to refrain from reading them, as they only spread lies and fantasies, but have no benefit.[48] Reading them does not make one smarter, wiser, more virtuous, or more pious. When worldly knowledge is compared to religious knowledge, the representatives of the latter meet with rejection in the world, but with approval from God.[49] Since this is a contrast, one side strengthens to the extent that the other is weakened: "Más vale una gota del divino saber, que los altos y profundos piélagos de la sabiduría mundana."[50] Worldly knowledge should therefore be shed in the interest of the divine. Finally, however, there are gradations in knowledge. The sciences that deal with external things are less valuable than the knowledge of one's own self: "Aunque sepas lo alto del cielo, y el profundo del mar, y todas las ciencias, si a ti no te conoces, eres semejante al que edifica casa sin cimiento."[51] However, interest in one's own self should not go too far and lead to excessive self-love that makes one forget everything else.[52] Thus, not only external things but also self-love are dangers in the search for truth.

In the mysticism of Miguel de Molinos, the negation of one's own self is even the condition for reaching the truth.[53] There are two basic principles to follow. First, one should stop considering oneself and whatever else there is in the world as particularly important. Then one should turn to God "para amarle, adorarle y seguirle sin género de interés propio".[54] Practice is more

---

[46] The School of Salamanca was particularly oriented towards Thomas Aquinas. Ramis 2024, 43.

[47] Estella 1980, 79.

[48] Estella 1980, 176.

[49] Estella 1980, 220.

[50] Estella 1980, 222.

[51] Estella 1980, 78.

[52] "los hombres que aman mucho a sí mismos, andan en grandes errores y no son alumbrados con el sol de la Sabiduría divina, sino encendidos con el fuego del propio amor." Estella 1980, 339.

[53] "El que no procura la total negación de sí mismo, no será verdaderamente abstraído, y así nunca será capaz de las verdades y luces del espíritu." Molinos 1976, 357.

[54] Molinos 1976, 361.

important than theory.[55] The revaluation of knowledge is summed up by the quote from Paul, which de Molinos translates and comments on: Let no one deceive himself. If anyone among you thinks he is wise in this world, let him become a fool to become wise. For the wisdom of this world is foolishness before God.[56]

The mystic Francisco de Osuna recommends inner composure and rejects external distraction. Changing locations to explore new things seems unnecessary. One should avoid moving and changing residence, as they only disturb inner composure. After all, it's not about the place, but about the inner life.[57] It is best to stay in one's study, where one does not stray from the path, just like in a safe ship.[58]

It has thus been shown that in the cited authors of religious literature following Augustine, some dogmatic determinations are made that control curiosity and postulate disinterest. One should engage with the eternal and divine, where practice is more important than theory and staying in one's own study appears particularly beneficial. Disinterest is required towards the world, worldly books, and all changes and innovations that one could experience on travels, as here lies not knowledge, but only ignorance. This dogmatic decision is conditioned by the pursuit of virtue, happiness, and eternal life through redemption. In the pursuit of knowledge, the following priority list emerges: Not the world, but God should be the object of knowledge, not the world, but the self should be the object of knowledge. Instead of dealing with the non-essential, one should deal with the essential, instead of dealing with many things, one should deal with the few, and instead of dealing with new things, one should deal with the tried and tested. Practice should always be preferred over theory, as the former is more directly connected with the self.

Some of these dogmatic knowledge-guiding pre-decisions are not new and can already be found in antiquity. They include that the self has priority over the world, that it is pointless to deal with non-essential objects, and that mere diversity and quantity do not represent knowledge gain. The Stoic Seneca associates curiosity with pleasure-seeking and considers it immoderate if someone w[59]ants to know more than is enough. A consequence of the Greek

---

[55] "Es regla general, y aun máxima en la mística Teología, que primero se ha de alcanzar la práctica que la teórica; primero se ha de experimentar el ejercicio de la sobrenatural contemplación, que inquirir el conocimiento e investigar la plena noticia de aquella divina ciencia." Molinos 1976, 356–357.

[56] Molinos 1976, 356 (Paulus 1, Ad Corint. 3, 18).

[57] Osuna 1972, 317.

[58] Molinos 1976, 320–321.

[59] However, the sentence could be interpreted on the one hand as a call to recognize the greatness of man thanks to his immortal soul, on the other hand to call for self-knowledge of individual weaknesses.
Tränkle 1985, 25, 28.

postulate "*Gnothi se auton!*", which was inscribed on the Apollo temple in Delphi, is for the humanist Erasmus of Rotterdam the focus on the own self, virtue, and human existence, whereby changes of location and travels can be seen as unrest and distraction.

In 17th century France, it is the mathematician, physicist, and author of literary works Blaise Pascal (1623–1662) who continues the criticism of pleasures and entertaining distractions. He advises on oneself. Close to the Jansenist movement, influenced by Augustinian ideas, he considers all activities that distract from the essential as diversions. Among the distractions, *divertissements*, which he rejects, the theater deserves the greatest condemnation.[60] He considers it wrong not to be able to leave one's own room, "de ne savoir pas demeurer en repos dans une chambre."[61] In distractions, he sees on the one hand a relief from the clear view of the lamentable situation of man, but on the other hand precisely the greatest misery of man.[62]

Eternity is more important than transience, as stated in Patristics. However, even antiquity gave priority to the constant over the inconstant, to the rules over the individual cases. Aristotle, and with him the second Scholasticism, refer to the wise as those who recognize difficult things that lie beyond sensory perception, as they search for the first causes and principles, preferring the superior to the subordinate science. This paradigm seems to be partially modified in the early modern period. The wonderful individual cases and exceptions now seem to arouse greater interest than the general and constant facts. The humanistic author Pedro Mexía (1497–1551) makes this his program.[63] He finds it wonderful in his essay-like *Silva* (1540) to consider the different contexts and inclinations of people with their special characteristics.[64] The exception, that which contradicts the ordinary rule, thus becomes attractive. Diversity and details are what attract attention and admiration.[65] This was also the view in the French 17th century. Descartes succinctly states that the extraordinary attracts attention: "Car nous n'admirons que ce qui nous paroist rare & extraordinaire."[66] Rare and extraordinary is also preferred by his contemporary La Bruyère: "La curiosité n'est pas un goût pour ce qui est bon ou ce qui est beau, mais pour ce qui est rare, unique."[67]

Nevertheless, curiosity should not be boundless. Warnings against useless knowledge are not uncommon in the Spanish *Siglo de Oro*. Even Pedro Mexía, who presents the most diverse and varied objects in his *Silva*, warns of

---

[60] Pascal 1976, 317.

[61] Pascal 1976, 93.

[62] Pascal 1976, 39.

[63] Mexía 1989b, 606.

[64] Mexía 1989b, 406.

[65] Mexía 1989b, 504.

[66] Descartes 1996, 116.

[67] La Bruyère 1964, 393.

the danger of dealing with worthless knowledge, which has increased due to printing, as useless books with lying stories appeared, which tired the minds and thus kept them from more valuable readings.[68] From the Latin word *cur*, the lexicographer Sebastián de Covarrubias derives the Spanish word *curioso* in his *Tesoro de la Lengua Castellana o Española* (1611). Therefore, the curious always ask why one thing is so and the other is different. Since too many questions not only testify to curiosity but also to idleness, the Spartans punished those who asked about things that were none of their business.[69] The derivation is indeed revealing, but wrong, since *curiositas* is derived from *cura*, which puts it in the context of care, concern, and effort that one brings to a thing or person. It thus steps out of the purely intellectual framework and points to a personal involvement and emotional commitment.

The concept of curiosity can be compared to that of interest. Especially with the word "interest", which consists of *inter* and *esse* and thus refers to being in between and being in it, participation becomes particularly clear. It has been pointed out that the word has undergone a special development in Spanish, where it has been used since the 15th century to distance oneself from materially successful Jews and Moors. The word *interés* is understood in this sense in religiously ascetic literature, for example in the *Tractado de República* (1521) by Alonso de Castrillo[70], as self-interest, selfishness, and egoism. Even in the French encyclopedia of the 18th century, the first meaning given for *intérêt* is that it is a vice that leads to enforcing one's own advantages at the expense of justice and virtue.[71] *Intérêt*, it is emphasized, is used by French moralists as a synonym for *amour-propre*. *L'amour-propre* is justified where it serves self-preservation, and to be rejected where it leads to disregard for others. Furthermore, "interest" also appears in relation to a literary work in the French encyclopedia as the turning of attention and emotional openness: "L'affection de l'ame qui lui est chere, & qui l'attache à son objet. [...] c'est l'attrait de l'émotion qu'il nous cause, ou le plaisir que nous éprouvons à en être émus de curiosité, d'inquiétude, de crainte, de pitié, d'admiration, & c."[72] And finally, the encyclopedia also mentions the economic meaning of *intérêt* as interest: "L'intérêt est le profit que tire le créancier du prêt de son argent."[73] In all the cases mentioned, the reference back to the self becomes clear, whether through personal benefit as in the first and last case, or through emotional participation in a literary work. Interest is therefore self-oriented and seeks its profit. Since interest as *amor sui* is seen as the opposite of *amor Dei*, the concept of *desinterés* as an expression

---

[68] Mexía 1989b, 22–23.

[69] Covarrubias, 2006, "curioso".

[70] Alonso de Castrillo, Tractado de república, Madrid 1958, 196.

[71] Diderot 1781, 888.

[72] Diderot 1781, 890.

[73] Diderot 1781, 893.

of selflessness emerged in Spain in the 17th century.[74] The love of God itself should no longer be practiced as *amor interesado*, a faith in the hope of redemption in the hereafter, but as *amor desinteresado*, for its own sake. This is what is meant in *Don Quixote* when Don Quixote cites knighthood as an example and Sancho ties it to disinterested love of God: "›What a foolish and simple creature you are, Sancho!‹ replied Don Quixote; ›do you not know that it is a great honor in our style of chivalry for a lady to have many wandering knights serve her, who have no other desire than to serve her for her own sake and ask no other reward for their enduring striving than the permission to be her knight?‹ 'That's exactly how I've heard it preached,' said Sancho Pansa; 'we should also love our Lord God for his own sake, without thinking of heaven and hell.'"[75]

Profitable for the protagonist in Miguel de Cervantes' *Licenciado vidriera* is the interest of others. He is modeled after the figure of the knowledgeable, who can competently provide information about everything and is therefore generally known and particularly respected.[76] The narrative can be divided into two parts: In the first part, the protagonist travels curiously through Europe, in the second he answers curious questions from others. His goal is to become famous through knowledge. Initially, however, he serves young gentlemen in Salamanca for eight years, which allows him to study law and humanities at the same time. Then he joins soldiers in the belief that extensive travels educate the traveler.[77] He experiences the hardships of sea voyages with storms and bugs, and marvels at the beauty of the city of Genoa. Wonder becomes his basic attitude in the face of the new countries and their inhabitants. After visiting Florence and Lucca, he marvels at the size of the city of Rome, admiring the bridges and hills, and his wonder grows when he sees Naples after another sea voyage. Only after he has marveled at Antwerp in Flanders does he return to Salamanca and acquire the title of licentiate of both rights. When he wants to get to know a lady out of curiosity, whose love he however rejects, she tries to change his mind with a magic potion. However, this has the unintended effect that he imagines himself to be made of fragile glass. As a result, he screams in fear when someone comes too close to him, he wears a noticeably wide garment and no shoes, adopts strange eating habits, sleeps in a field in summer and in the barn of an inn in winter.

---

[74] Corominas 1961, 331.

[75] "Oh qué necio y qué simple que eres!—dijo don Quijote—Tú no ves, Sancho, que eso todo redunda en su mayor ensalzamiento? Porque has de saber que en este nuestro estilo de caballería es gran honor tener una dama muchos caballeros andantes que la sirvan, sin que se estiendan más sus pensamientos que a servirla por solo ser ella quien es, sin esperar otro premio de sus muchos y buenos deseos sino que ella se contente de acetarlos por sus caballeros.—Con est manera de amor—dijo Sancho—he oido yo predicar que se ha de amar a Nuestro Señor, por sí solo, sin que nos mueva esperanza de gloria o temor de pena." Cervantes 1998, 363–364.

[76] Cervantes 1982, 117.

[77] Cervantes 1982, 107.

Since mind and soul penetrate glass faster than a heavy body, he believes he can give the right answers to all possible questions. In the second part, it is the scholars of the university and all professors of medicine and philosophy who are amazed at his sharpness of mind. He evokes "admiración y lástima"[78] in everyone he meets. So curiosity is now on the side of the questioners. It refers not only to the objects of the questions, but to the Licenciado himself.

What kind of questions are being asked? The questions usually relate either to situations of everyday life or to common professions. Rarely is knowledge asked for, more often its representatives. An exception is poetry, of which the Licenciado says that it includes all other sciences, uses them, and produces wonders, with which it enriches the world usefully and entertainingly.[79] Such serious answers testify to wisdom, while the playful results are of satire or self-irony. An example of self-irony is the answer to the question why he could feel pain after a wasp sting on the neck, since he was made of glass. He replies that the wasp must belong to the guild of slanderers, whose tongues and stings are sharp enough to penetrate a bronze body. The question of how the man whose wife has run away with another should behave, on the other hand, is answered by the wise and practical advice that he should thank God for having his enemy abducted from the house. Also the question of how the man should live in peace with his wife leads to well-meant advice: She should command over all people in the house and he over her. On the other hand, the advice given to someone who asks what they can do to avoid envying others is witty but not very promising: They should sleep, because then at least in their sleep they will be equal to those they envy. And at the expense of the curious questioner, the reader is amused by the witty, but practically useless answer to the question of who is the happiest person: *Nemo; porque Nemo novit Patrem, Nemo sine crimine vivit; Nemo sua sorte contentus, Nemo ascendit in coelum.*[80]

Even where questions are asked about professions and their characteristics, the answer is satirical and without claim to truth. In response to the question of why poets are usually poor, it is said that it is their own business, as it would be easy for them to be rich and to avail themselves of the golden hair of their ladies, their foreheads of polished silver, eyes of green emeralds, lips of coral or the liquid pearls of their tears. When asked about the mistake of booksellers, he explains how they cheat writers out of their wages. In response to the sedan carrier's question of what he thinks of him, he replies that every sedan carrier knows more sins than a confessor, but unlike the latter, does not keep them to himself, but hears them to recount them in the taverns. About doctors, he comments that they are the only ones who can take our lives without having to fear punishment. He compares lawyers to

---

[78] Cervantes 1982, 118.

[79] Cervantes 1982, 122.

[80] Cervantes 1982, 135.

doctors, who always keep their sheep dry, whether the patient recovers or not. So, it is usually something negative that characterizes the profession. In this way, he satirically and skillfully comments on donkey drivers, sailors, cattle drivers, pharmacists, fencing masters, chaperones, scribes, slanderers, gambling house owners, investigating judges, false scholars, tailors, shoemakers, bakers, puppeteers, and actors.

When the Licenciado is finally cured and no longer considers himself a glass man, people still follow him out of curiosity, so that he cannot practice his profession as a lawyer. The persistent curiosity of the crowds constantly gathering around him drives him to ruin, so he returns to Flanders, where he eventually dies as a wise and brave soldier.

The travels of the Licenciado, like those of Odysseus, allow him to have previously unknown experiences. The Licenciado's repeated reaction is to wonder. Considering that, according to the Licenciado, literature, as quoted, offers not only utility and pleasure, as has been customary since Horace, but also wonder, then wondering seems to be another keyword of the novella. What is the meaning of the word in the *Siglo de Oro*? In Pedro Mexía's widely read *Silva de varia lección*, the wonderful is met with the attitude of admiration.[81] Wondering can according to Augustine lead to the worship of God through his wonderful works.[82] In Mexía, as in scholasticism, Aristotle is meant when he speaks of "the philosopher" and traces back to him the idea of the natural curiosity of man, which has even led to the exploration of the stars and planets.[83]

Indeed, Aristotle considers wondering to be the trigger for philosophizing. "For wonder was the beginning of philosophizing for men now as in the past, as they initially wondered about the most immediate inexplicable things, then gradually progressed and raised questions about larger things, such as the phenomena of the moon and the sun and the stars and the origin of the universe."[84] Finally, Aristotle's Metaphysics begins with the statement that all men by nature strive for knowledge. A sign of this is their love of sensory perceptions. All knowledge begins according to Aristotle "with the wonder that things are as they are, such as in the face of self-moving puppets, the solstice or the incommensurability of the diagonal."[85]

Thus, wonder is the prerequisite for new insights, and those who cannot wonder remain ignorant. At least this is the conclusion drawn by René Descartes from the Aristotelian sentence: "mais nous n'avons que l'admiration pour celles [les choses] qui paroissent seulement rares. Aussi voyons nous que ceux qui n'ont aucune inclination naturelle á cette passion, sont

---

[81] Mexía 1989b, 498; Mexía 1989b, 504.

[82] Augustine 2000, 293.

[83] Mexía 1989b, 177.

[84] Aristotle 1995a, 6.

[85] Aristotle 1995a, 23.

ordinairement fort ignorans."[86] Now it becomes understandable why the Licenciado receives the news he perceives on his travels with an attitude of wonder. The degree of his astonishment indicates the extent of his increase in knowledge.

Numerous are the parallels between the anonymous *Segunda Parte del Lazarillo* (1555)[87] and the *Licenciado vidriera* (1613). Both protagonists undergo a transformation, Lazarillo into a tuna and the Licenciado into glass, and a retransformation. Both appear so knowledgeable after numerous travels that they are asked more or less difficult questions. While Lazarillo owes his salvation by the tunas to excessive wine consumption, the Licenciado learns about the variety of wines after a dangerous sea voyage at a generous wine tasting. While the Licenciado is driven by a thirst for knowledge, it is greed that makes Lazarillo embark on his journeys.[88] Material things appear more important to Lazarillo than intellectual ones. Therefore, he chooses for his council twelve of the richest and not the wisest tunas.[89] This does not prevent Lazarillo from eventually being titled as "el más cuerdo y sabio atún que hay en el mar"[90]. Both curiosity and greed are reprehensible passions that greedily, thus in an exaggerated manner, strive for intellectual or material possession. In both cases, the measure necessary for virtue is lacking.

The rector of the university asks Lazarillo four questions: To the first, how many barrels of water are in the sea, he answers that if all the water could be held together in one container, he could measure it. The second question, how many days have passed since the creation of Adam to today, he answers with reference to the seven days of the week, to which seven days always follow. Already now, Lázaro appears as particularly clever among the clever doctors and as a teacher among the students.[91] The question of where the end of the world is, he counters with the counter-question, what kind of speculations these are. How could he answer, since he has never been there.[92] The last question about the distance between heaven and earth provokes protest, which is articulated in an inner monologue in which he complains that the questioner should know that he has not yet gone there. He should have asked him about the ways of life and the language of the tunas, which he has learned from his own experience.[93] He could therefore answer well if he were asked about the experiences he has gained, but not in relation to a path of

---

[86] Descartes 1996, 116.

[87] Anónimo 1988.

[88] "Con esto y con la codicia que yo me tenía, determiné—que no debiera—ir a este viaje." Anónimo 1988, 131.

[89] "Yo escogí para mi consejo doce dellos, los más ricos, y no tuve respeto a más sabios si eran pobres." Anónimo 1988, 179.

[90] Anónimo 1988, 212.

[91] Anónimo 1988, 254.

[92] Anónimo 1988, 255.

[93] Anónimo 1988, 255.

which the questioner should know that he has not yet taken. Thus, he claims that heaven is very close to earth, and proposes a practical experiment: The rector should go to heaven, then he, Lazarillo, would hear him sing, even if he only sings softly. From now on, Lazarillo enjoys similar popularity as the Licenciado. For all present find his answers excellent, want to congratulate him, see him and hear him speak. His name is on everyone's lips: "El nombre de Lázaro estaba en la boca de todos, y iba por toda la ciudad con mayor zumbido que entre los atunes."[94]

It becomes clear that the rector's questions relate to speculative subject areas that are not accessible to experience. However, it is experiential knowledge and the practice of experience that Lazarillo bases his answers on, which is why he does not actually answer the questions, but clarifies that they cannot be answered and are irrelevant for the practical life of humans, i.e., that they originate from misguided curiosity. When the rogue Lazarillo presents himself as wise in his answers, he is again comparable to the Licenciado, who presents himself as wise with his madness, *locura*. Both thus correspond to the pattern of the wise fool, to which Jerónimo de Mondragón dedicated himself in his *Censura de la locura humana y excelencias della* (1598) in reference to *Encomium moriae* (1510) by Erasmus of Rotterdam.[95] Mondragón cites the example of a man who considered all the ships that docked in his city's harbor as his own. After his recovery from this folly, he insisted that he had never been as happy as with his folly. The Licenciado also fares better before his recovery. A similar dialectic is evident in Mondragón's thesis that the foolish crowd can only applaud a false sage, while it mocks the true one and declares him a fool. If this thought is applied to Lazarillo or the Licenciado, there is the possibility that the audience, foolish in their own way, venerates them as knowledgeable when they are in reality ignorant. It is also possible, and this is closer to the irredeemability of the audience's expectations, that the foolish audience only recognizes a pseudo-knowledge, but fails to recognize the wisdom hidden behind it, which is expressed in the critical attitude and playful manner of the replies of both.

The concept of world curiosity is considered a keyword for the departure from medieval thinking and living orders.[96] First, the conflict between happiness and curiosity should be clarified, which each require dealing with different objects.[97] Seneca criticizes in *De brevitate vitae* the curiosity that asks unnecessary questions of unnecessary scholarship, and scolds the Greeks who wanted to determine how many rowers Odysseus had, and whether the *Iliad* or the *Odyssey* was written first. The new experiences that can be gained on travels are numerous, Seneca adds, but do not make the traveler better or

---

[94] Anónimo 1988, 257.

[95] Strosetzki 1987, 118–119.

[96] Blumenberg 1973.

[97] See Joly 1961, 13–44 in the following.

stronger (*neque meliorem*, [...] *neque saniorem*). Here Seneca's stoic attitude is evident, according to which it is more important to know the central and general principles of ethics, logic, and physics than the diversity of individual people and things. And if the Stoics want to convey ataraxia, i.e., independence from the world, this is not done through science. Happiness in life is valued higher than truth of knowledge. Cicero had rejected in his writing on the state unnecessary and fruitless investigations about a second sun, which could do no harm, as these served the attainment of neither happiness nor morality. Homer had, Cicero thinks, rightly seen that the Sirens had to promise Odysseus something special, namely science, to seduce him. Nevertheless, curiosity about everything and everyone is to be condemned, while it is to be praised when directed at essential things. This contrasts a negative type of curiosity with a positive one. Cicero sees his position already laid out in Greek Epicureanism, which the study of music, geometry, arithmetic, and astronomy, even if it led to true statements, could not help to live better and happier. Epicurus holds curiosity responsible for the affects of fear and hope, which destroy human happiness when they dominate. Polymathy is already considered useless by the pre-Socratic philosopher Heraclitus, as mere factual knowledge does not form the mind. The critical attitude towards *curiositas* among Christian authors like Augustine thus results from the takeover of an ancient, especially stoic attitude.[98]

Augustine accuses the astronomer of excessive curiosity about the mysteries of nature in his *Confessions*. He condemns those who count the stars and the grains of sand on the beach out of curiosity, measure celestial regions and calculate star paths, and can predict solar and lunar eclipses by day, hour, and location. "They see future solar eclipses long in advance, but they do not see their own present one. [...] They say much that is true about creation, but they do not seek the truth, the artful designer of creation."[99] Augustine also speaks of the contemptibly small things that tempt our curiosity every day. We first listen to empty chatter out of politeness, then out of interest. A dog chasing a hare in the circus is of no interest. But in the open field, this hunt arouses curiosity and perhaps distracts from a great thought. The same can happen when curiously observing a lizard catching flies, or a spider ensnaring them with its webs.[100] To such an argument, Mexía counters that moral lessons for human life can be derived from observing animals.[101] Similarly, Thomas Aquinas rejects curiosity that stops at the transience of individual things and does not refer them back to their origin in God as creation, even

---

[98] Joly 1961, 44.

[99] Augustine 2000, 118–119.

[100] Augustine 2000, 292–293.

[101] Not only "avisos para [la vida y salud, pero reglas y ejemplos para] las virtudes y buenas costumbres." Mexía 1989c, 187.

though he, like Aristotle, considers the pursuit of knowledge to be natural.[102] The Spanish humanist Mexía comments on investigations as to why snow covered with straw remains cold and warm water covered with straw remains warm, the results of which satisfy the mind but are otherwise worthless.[103] The danger of dealing with worthless knowledge has increased especially due to the printing press.[104]

And Pérez de Oliva explains that wisdom remains closed to the essentially straightforward human being if his will occupies itself unnaturally and maliciously with superfluous questions.[105] In reference to the Greek demand that one should first know oneself, the humanist Erasmus demands a focus on the self, virtue, and human existence. He rejects the curious gaze here and there as well as the dispersion and unending, dangerously acquired individual knowledge. The German word "Erfahrung", taken literally, has to do with travel, i.e., changing location. In this context, travel appears as unrest, which distracts from spiritual exercises, in the Middle Ages as a sign of *acedia*, i.e., as neglected care for personal salvation due to laziness. In its wake, vices such as gluttony and sensuality set in. The prototype of the peace-disturbing wanderer is Cain; his positive counterpart bears traits of the cloistered and home scholar. Thus, the unique and wondrous curiosities collected on the voyages of discovery of the New World were carefully ordered according to Aristotelian categories. Such curiosities are also listed by Mexía in his *Silva*, not without pointing out their uselessness and insignificance.[106]

Astonishment, horror, or curiosity are passions of cognition, which usually presuppose an object that violates the familiar, an anomaly. Therefore, one can argue[107] that the medieval scholastics, starting from Aristotle, were primarily concerned with what always or mostly happens in nature, while some early modern authors set out to explore the wonders and hidden secrets of nature. Thus, because wonders and secrets have become preferred objects, astonishment and curiosity are also upgraded as philosophical passions. Indeed, the preference for the general over the concrete goes back to Aristotle, who in his *Metaphysics* recognizes as wise the one who knows everything without having the science of the individual. According to Aristotle, he can recognize difficult things that lie beyond simple sensory perception. The wise person seeks the first causes and principles, preferring the superior science to the subordinate one. The fact that in the 16th century, alongside the preoccupation with the normal, astonishment at the

---

[102] Blumenberg 1973, 132.

[103] Mexía 1989b, 142.

[104] "Books of little fruit and profit, of fables and lies," "destroy and tire the minds and divert them from good and healthy reading and study." Mexía 1989c, 22–23.

[105] Pérez de Oliva 1967, 119.

[106] Mexía 1989a, 332.

[107] Daston 2002, 161.

extraordinary also emerges, is also confirmed by the humanist Mexía, who finds most interesting what collides with usual thinking habits.[108]

However, the fact that the individual new has always been preferred as the subject of literature, because it could hope for the interest and curiosity of the reader, somewhat relativizes the importance of novelty a little. Mexía cites Plutarch's parallel biographies with "grandes y notables exemplos, que podrá ver allí el amigo y curioso de hystorias".[109] Lazarillo also wants to present unheard-of things and, like Apuleius, resorts to the topos "I bring something never said before", which can already be found in the context of the opening topic at the end of the 5th century BC.[110]

While the definitions of curiosity mentioned so far start from its objects, the following will consider the subject, i.e., the state of the curious person. Indeed, in the early modern period, the dangers of knowledge were primarily seen in the motivation of the subject, e.g., in curiosity, while today they are mainly seen in its application, e.g., in biotechnology. Since, according to the Church Father Lactantius, God has hidden what serves the satisfaction of mere curiosity[111], the limits of human reason appear violated by the pursuit of experience where none can exist. The arrogance towards God, Greek *hubris*, can also be directed against the other members of a community and then manifests itself in dogmatism. Whoever is primarily concerned with the new, or with *invenienda*, instead of *inventa*, according to *De vana curiositate* of the theologian Jean Gerson (1363–1429), leaves the truth-guaranteeing consensus of scholars, the "concordantia doctorum".[112] The accompanying attitude is empty vanity (*vanitas*) and not piety (*pietas*).

The connection between curiosity and cupidity is already evident in idioms, such as those used by Mexía, who considers people to be greedy for knowledge, "cobdiciosos de saber"[113], or by Villalón, who speaks of a curiosity for wealth, a "curiosidad de adquerir riquezas".[114] The ancient author Plutarch characterizes in his treatise *De curiositate* adultery as a kind of curiosity about another's pleasure.[115] Augustine equates the addiction of curiosity with carnal desire. He calls it "the lust of the eyes", as it is based on the desire for knowledge and the sensory organ of the eye plays the main role in recognition. Sensations in plays appeal to this greed. They move people to seek secrets, to turn to magic, and to demand signs and wonders, just

---

[108] "aquello que paresce que repugna al común ser y orden de las cosas." Mexía 1989b, 606.

[109] Mexía 1989a, 502–503.

[110] Curtius 1973, 95.

[111] Müller 1976, 314–315.

[112] Müller 1976, 317; Gerson 1962, 230, 238.

[113] Mexía 1989b, 177.

[114] Villalón 1967, 42.

[115] Blumenberg 1973, 90.

to experience something.[116] Experience is always subject to criticism when it serves vice and leads to worldliness. A vivid example is Cesare Ripa's pictorial representation of curiosity in his *Iconologia* (1593). She wears ears all over her garment to be able to hear everything, she looks with greedy frog eyes, and has raised hands and tousled hair to express her excitement.[117] Pedro Mexía advises avoiding the one who asks too much, as he is a chatterbox .[118]In La Bruyère, it is finally said for the courtly context of the 17th century: If its object were less insignificant, then the passion of curiosity would be of no less power than that of love or ambition.[119]

It has thus been shown that the encyclopedic attitude, which wants to absorb everything from the infinite amount of knowledge, is opposed by the dogmatic one, which rejects the former as eclectic and rejects everything else from a fixed standpoint. The former attitude had its predecessor in the universal and polyhistorical knowledge that rhetoric demanded from the good speaker. Dogmatism is evident when the philologist dismisses the knowledge of the non-philologist as pedantic, when the natural scientists contrast their exact and useful knowledge with the fantasy constructs of others, or when the scholastics label representatives of other directions as deviants. In Augustine, the highest should be the object of wisdom, to which everything changeable should be met with disinterest. In Diego de Estella, it is the priority of the eternal that makes worldly knowledge and profane readings worthless. In Miguel de Molinos and Francisco de Osuna, this goes so far that the own self is negated and any kind of change of location should be avoided. The latter is joined by Pascal when he rejects diversions as *divertissements*, i.e., distractions from the essential.

While Aristotle and the second Scholasticism with him preferred the constant over the variable and the rules over the sensually perceptible individual cases, a counter-movement emerges in the early modern period with Pedro Mexía and La Bruyère that favors wonderful individual cases and exceptions, which are supposed to particularly satisfy curiosity. Thus, curiosity becomes a disputed characteristic, which is associated with material interest and self-love and contrasted with selflessness. The questions and answers of Cervantes' *Licenciado vidriera* and Lazarillo attack curiosity through satire and make speculative knowledge appear useless compared to experience. Examples of curiosity for useless knowledge include questions about the number of rowers Odysseus had on his ship, or about the existence of a second sun. Attention easily turns into interest and curiosity. However, when it comes to knowing

---

[116] Augustine 2000, 290–291.

[117] Krüger 2002, 13.

[118] Mexía 1989a, 214.

[119] "Ce n'est pas un amusement, mais une passion, et souvent si violente, qu'elle ne cède à l'amour et à l'ambition que par la petitesse de son objet." Bruyère 1964, 393.

oneself above all, travels become disturbing distractions and curiosity turns into greed, dogmatism, and hubris.

2.4   DOUBTS ABOUT KNOWLEDGE

If one distinguishes the recognizing subject from the object to be recognized in cognition, the question arises as to how the object comes to the subject in cognition. It is philosophical skepticism that fundamentally denies this possibility by pointing out that the sensory perceptions that the subject has of the object world are unreliable. Thus, the very sensory experiences that form the basis of "positive knowledge" in the 19th century are questioned in the early modern period. To illustrate this, let's briefly look at the work *Que nada se sabe* by the physician and mathematician Francisco Sánchez, which appeared in Latin in 1581. For him, the real knowledge is an internal one, as the sensory perception of external things can deceive and is therefore uncertain.[120] The mind alone reaches the substance of things, while the senses often deceive it on this path. This is the case when distant things appear small and blurry and when things close up are not seen at all or as they are. What moves quickly far away appears slow to the senses. An example of this are the stars.[121] If sight and hearing together want to grasp the external space, the confusion becomes even greater. When a horse gallops, the listener has the impression that there are two. When two run in step, one hears only one.[122] That human perception can also depend on training is made clear by the skeptic Pedro de Valencia (1555–1620) when he emphasizes the importance of experience and practice for the artist. The painter sees numerous details in his pictures that escape the layman. Similarly, a singer hears nuances that remain hidden to others.[123] And how does the mind behave when it is repeatedly deceived by the external senses? According to Francisco Sánchez, small errors at the beginning become large ones at the end of the chain of thought. If something appears wrong to the mind, it looks beyond the external sensory information for the cause of the error. By orienting itself to probabilities, doubt is perpetuated to infinity.[124] Again, it is the thinking subject that has the greatest share when it comes to knowing something about the external things.

Montaigne's essay *Apology of Raimond Sebond* is oriented towards Sextus Empiricus' *Outlines of Pyrrhonian Skepticism*, which is why Montaigne was considered a skeptic by contemporary readers. In other essays as well, Montaigne thinks as a skeptic, once through the method of juxtaposing

---

[120] Sánchez 1991, 120.

[121] Sánchez 1991, 124.

[122] Sánchez 1991, 128.

[123] Valencia 2006, 5.

[124] Sánchez 1991, 130.

contradictory statements and then by emphasizing subjectivity.[125] The unsystematic form of the essays allows for contradictions to be placed side by side. It is Montaigne's opinions that matter: "Je peins principalement mes cogitations, subject informe, qui ne peut tomber en production ouvragere."[126] He cannot discuss things as they are, but only as they appear to him. The diversity of living beings results in different perceptions. Montaigne points to the sense of time in the rooster and the knowledge of some animals of healing herbs, which gives the animals a different perspective on the world. These hidden abilities that we see in animals exceed our possibilities. For humans too, the world would appear differently if one or two senses were missing. If experiences are already different, then the generalizations derived from experiences are even more so. After all, two people never judge the same thing in exactly the same way. It is impossible to find two completely identical opinions on the same thing, not only among different people, but also at different times for the same person.

It is clear that the genre corresponding to skepticism cannot be a treatise, which is why Montaigne titles his expositions *Essais*, i.e., attempts, subtly indicating that no definitive truths are to be expected. He also uses the terms *fantaisies, imaginations, ravasseries, rien que du rien,* and *caprice* and refers to the results of chance and the activities of his *niaiser* and *fantastiquer*.[127] His oft-quoted question "Que sçay-je?"[128] serves to recognize non-knowledge, which is why he likes to use negation and considers the objects themselves secondary to the manner of expression.[129]

Another example of skepticism in the early modern period is the Spanish Baroque author Baltasar Gracián. As is well known, he provides hints in his treatises that allow the reader to find the way to the perfect *político, héroe,* or *discreto*. The aphorisms of the *Oráculo manual* persue this goal, and the novel *Criticón* shows the practice of the pedagogical process in the form of an allegory. Gracián therefore seems to convey convictions rather than doubt them. However, it is conceivable that he approaches his work with skeptical distance. Thus, he could in principle doubt the completeness and truth of his insights, or their attainability by those to whom they are to be conveyed. And finally, dogmatism and skepticism are not incompatible. It is possible to hold on to opinions and yet deny the possibility of recognizing them as true with reason.

In the research literature, Gracián's skeptical elements have not yet been systematically appreciated. The similarities in the evaluation of *desengaño*, that is, the overcoming of a deception, in Arthur Schopenhauer and Gracián were

---

[125] Wild 2009, 109–133; Strosetzki 1982, 89–104.

[126] Montaigne 1962, 359.

[127] Westerwelle 2002, 200, 203, 205, 229.

[128] Montaigne 1962, 508.

[129] Westerwelle 2002, 209.

seen[130] and in Gracián, in addition to a practical orientation towards tactics, a theoretical-philosophical interest in truth was also highlighted.[131] There was also talk of the central importance of deception in the *Siglo de Oro*.[132] After all, Gracián says that things are not considered for what they are, but for what they seem. Therefore, it is all the more important to present them in the right light.[133] Skepticism was also found in interpersonal relationships, especially in the assessment of human ways of life, actions, and motives.[134]

The following will therefore take a further step and look for Gracián's answers to the question of whether it is possible to gain objectively valid knowledge. And, if so, whether it is possible to pass on any insights completely and unambiguously. The first starting point is the criticism of sensory perceptions, and an important target is the criticism of ethical norm settings. Diogenes Laertius[135] has already listed ten skeptical tropes, i.e., reasons for the impossibility of knowing the truth, in his chapter on the Greek skeptic Pyrrhon of Elis: Humans cannot claim true knowledge for themselves, as all living beings are different in terms of their senses, pleasure, pain, harm, and benefit. For example, falcons have sharper eyes, dogs have a sharper sense of smell. Differences in visual acuity result in different appearances. In terms of physical constitution, there are different human natures as starting points. Since the apple is yellow to the eye and sweet to the taste, different sensory impressions are associated with each object. Moods or changes in states such as illness, health, joy, suffering, youth, old age, courage, fear, but also views of good and evil, of beauty and ugliness influence knowledge. Spatial relationships, place and time, habit and novelty, speed and diversity of things change their knowledge. Whether and to what extent such thought processes play a role in Gracián will be shown in the following.

Skepticism is by no means alien to theology. Following the skeptical approach, since the Middle Ages, Christian theologians have acted as agnostics for the double truth, have a different understanding and interpretation of something, depending on whether it is viewed from the standpoint of revelation or reason, i.e., theologically or philosophically. They also speak of the *deus absconditus* and reject metaphysics as unknowable. It is the Jesuits who defend the view with probabilism that one cannot reach certainty in science and philosophy, but must be content with more or less great probabilities. In case of doubts about the permissibility of an action, the less certain opinion,

---

[130] Neumeister 1991, 261–277.

[131] Neumeister 1993, vol. II, 735–739.

[132] Schulte 1969, 77–91; Strolle 1976, 5–17.

[133] Gracián 1986, 411; "Las cosas no pasan por lo que son, sino por lo que parecen; son raros los que miran por dentro y muchos los que se pagan de lo aparente." Gracián 1986, 399; "Es el Engaño muy superficial y topan luego con él los que lo son. El Acierto vive retirado a su interior, para ser más estimado de sus sabios y discretos." Gracián 1986, 211.

[134] Jansen 1958, 211.

[135] Diogenes Laertius 1950, vol. II, 492–497.

but supported by probable reasons, may be followed even if an opposing one would be more praiseworthy.

In the philosophy of the early modern period, skeptical approaches are common. The already mentioned Frenchman Montaigne is not only an agnostic and denies the possibility of metaphysical knowledge, but also doubts the abilities of reason and the senses. His compatriot Descartes will develop from this a century later the methodical doubt, which recognizes the only basis for the construction of truth in the doubting self. His contemporary Gracián is comparable to Descartes insofar as he prefers practice and experience to the authorities of antiquity.[136] He also adopts numerous skeptical tropes that we have presented following Diogenes Laertius.

However, it should not be overlooked that the philosophy of skepticism already had prominent representatives in Spain in the 16th century. An example is Pedro de Valencia.[137] Less well-known than the science critic Vives[138] are the doctors and skeptics Gómez Pereira with *Antoniana Margarita* (1554)[139] and Francisco Sánchez with his work *De multum nobili et prima universali scientia. Quod nihil scitur* from the year 1581.[140] Both anticipate in their own way the doubt of Descartes, which is why they were often compared with Descartes.[141] Since Gracián, as will be shown, also presents Cartesian elements, they can also be considered his precursors.

In the following, the principal obstacles that, according to Gracián, make knowledge difficult will be presented in a first step. Then the problems of recognizing one's own subject and the object will be addressed. This is followed by Gracián's assessment of the actions of the various subjects with each other and his evaluation of the necessity or possibility of true knowledge of the respective counterpart. Finally, Gracián's assessment of the possibilities and limits of truth transmission and education will be asked.

The principal difficulties include firstly the diversity and changeability of things, which make the possibility of generalization to a unity appear problematic. Secondly, it also includes that different knowledge is produced depending on the respective interests or affects. If there are thirdly different degrees of certainty of knowledge, then the question arises at which stage of certainty truth can be spoken of. Fourthly, Gracián critically examines how knowledge is usually gained.

If Gracián repeatedly points out the diversity of things and the many aspects under which they are to be considered, this could be interpreted as a call to recognize unity beyond diversity or to present the knowledge of unity

---

[136] Gracián 1967, 877–892.

[137] Gómez Canseco 1993, 105–130.

[138] Strosetzki 1995.

[139] Wigger 2001, 262–279.

[140] Sánchez 1991.

[141] On Gómez Pereira cf. Wigger 2001, 263; on Sánchez cf. Iriarte 1935; Moreau 1960, 24–50.

as impossible. One can opt for the former and recognize a certain unity in Gracián's perspectivism, not relativism.[142] This can be interpreted in terms of the lifeworld.[143] Or, with reference to the doctrine of double truth, two separate paths to a great goal can be recognized.[144] However, what is missing is an attempt to clarify whether the unity or the goal is a "thing in itself" or an individually subjective construction independent of the "thing in itself". From Gracián's assertion that everything happens in images and imagination in this life, and is appearance, which cannot be touched with hands[145], it can be concluded that the diversity of sensory representations is processed into images with the imagination, which are taken for knowledge, but do not reach the things in themselves.[146] Accordingly, the moral or aesthetic evaluation of objects is not founded in the object, but in the subject, as Protagoras already said: "Protágoras decía que en las cosas no había bien ni mal, pesar ni gusto, sino en la imaginación y en el modo de concebir de cada uno."[147]

Elsewhere, however, Gracián seems less to anticipate late Kantian positions, but rather to draw on the Aristotelian distinction between substance and accident. When he distinguishes the *Veedor de todo* from ordinary people by the fact that the former always grasps the substance along with the accidents, he emphasizes the construction of an absolute knowledge from that which is humanly possible. In contrast to the substance, the first of the Aristotelian categories, which is thought of as the enduring essence of an object, the accidents are what befalls an object and what can be said about it. However, they are only what is not necessary and not usually befalling it, thus the accidental, changing, non-essential properties and states of an object. Thus, the *Veedor de todo* claims of himself that he sees the substance of things in an instant, not just the accidents like humans.[148] Thus, absolute knowledge is only possible for an absolute being and not for humans.[149] The diversity and infinity of changing forms of appearance prevent the art of deciphering everything from being fully learnable for humans.[150]

The diversity on the side of the object corresponds to the diversity on the side of the subject, which seems to reflect the macrocosm of the object world as a microcosm in miniature.[151] So if the diversity of objects making

---

[142] Moraleja Juárez 1999, 93.
[143] Heger 1982, 68.
[144] Krauss 1947, 157.
[145] Gracián 1967, 738.
[146] Schröder 1966, 83.
[147] Cf. Schröder 1966, 93.
[148] Gracián 1967, 897.
[149] Gracián 1986, 124.
[150] Gracián 1967, 880.
[151] Gracián 1986, 264.

an impression on the subject is added to a diversity of ideas on the side of the subject, then knowledge is hindered on both sides.

Gracián provides suitable examples for the unreliability of the subject and begins, like Descartes' methodical doubt, with the criticism of sensory perception: The subjective perception of one person differs from that of another. What appears white to one is black to others.[152] But also fundamentally similar sensory perceptions can be interpreted differently and lead to contradictory ideas of reality. When Critilo and Andrenio meet a man of strange appearance, one perceives that he is going, the other that he is coming.[153] In view of this diversity of ideas about reality, according to Gracián, it would only be a testimony of ignorance if one were to insist on one's own opinion. A healthy measure of doubt and flexibility seems appropriate in view of the general uncertainty.[154] Where convictions are as numerous as tastes or faces, according to Gracián, there is a general relativism that makes the exclusive claim of truth the subject of laughter and makes truth intersubjectively dependent on voices and opinions.[155]

An important factor that causes change and diversity on the side of the subject is time. Since humans are in constant change, their thinking and thus the basis of knowledge changes with their development: Over the years, humans change. At twenty, he is a peacock, at thirty a lion, at forty a camel, at fifty a snake, at sixty a dog, at seventy a monkey, at eighty nothing.[156] Changes are not only caused by age, but also occur in shorter periods. What was presented as true yesterday is untrue today. There are people who are different every day and have different opinions.[157] Finally, it is fashions and habits that condition thinking and decision-making so much that the one who persists in old ways of thinking and does not follow the latest fashion appears ignorant.[158]

However, not only the subjective and objective diversity caused by time thwart knowledge, but also the impulses that have triggered the process of knowledge. Recognition as an action is triggered by the will. Like Descartes, Gracián also sees the reason for error in the fact that the will to recognize something is greater than the ability of the mind to do so. If the mind is not advanced and experienced enough, it should not risk error, but stay where it is.[159] Equally misguided is the well-trained mind when it is guided by a bad

---

[152] Gracián 1967, 681.

[153] Gracián 1967, 835.

[154] Gracián 1986, 433.

[155] Gracián 1986, 399.

[156] Gracián 1986, 469.

[157] Gracián 1986, 388; cf. Gracián 1986, 263.

[158] Gracián 1986, 407.

[159] Gracián 1986, 124.

will. According to Gracián, it leads to violence.[160] Those who are driven by mere will and not by judgment and prudence do not arrive at truth, but at confusion.[161] On the other hand, one cannot do without the will. So it is not enough to convince the mind alone if one does not at the same time win the will of the other.[162] The will is as beneficial as it is dangerous.

Closely related to the will is the interest in knowledge. These are impulses not guided by reason, for which reason then only provides justifications.[163] Also, other factors that cloud the clarity of reason according to Gracián, such as passions, can be cited.[164] Indeed, Gracián clearly shows how these can guide observation, for example.[165] Accordingly, reason is also misguided as often as it entrusts itself to the guidance of the affections[166] or the capriciousness leading to different decisions every day.[167]

Pure reason, unclouded and uninfluenced, should therefore be the ideal instrument of knowledge, one might conclude. But here too, Gracián raises reservations and emphasizes that practice and prudence alone are not enough. Therefore, taste and judgment must be added to them.[168] So that neither study nor wit are enough where choice is lacking.[169] According to Gracián, unclouded understanding and isolated knowledge are insufficient. On the Island of the Immortals in the *Criticón*, only those are admitted who have enriched their theoretical knowledge with experience and who mitigate the fundamental relativity of all knowledge by possessing the art of wisdom and applying it to knowledge.[170]

If someone does not know at all what to choose as the truth from the diversity of reality, then it makes sense for him to protect himself from mistakes by always choosing the option that seems safest to him from different possibilities.[171] Even if he then only orientates himself towards probability and is not considered particularly clever, he will still be considered solid. If one does not know which opinion to join, then one should join those who one believes are in possession of knowledge.[172] The powerful have it good,

---

[160] Gracián 1986, 366.
[161] Gracián 1986, 116.
[162] Gracián 1986, 138.
[163] Gracián 1986, 476.
[164] Gambin 1993, 65.
[165] Gracián 1986, 291.
[166] Gracián 1986, 124.
[167] Gracián 1986, 298.
[168] Gracián 1986, 380.
[169] Gracián 1986, 279.
[170] Egido 2001, 21.
[171] Gracián 1986, 467.
[172] Gracián 1986, 372.

accustomed as they are to compensating for their own ignorance with the knowledge of others.[173]

Since it is already necessary, in cases of uncertainty, to forego one's own knowledge and to align with the judgment of those who appear competent, the question arises whether knowledge is usually acquired in a different way. And here Gracián gives a surprisingly modern-seeming answer and states that information usually does not come from perception or personal experience, but second-hand, and what is true is what others consider to be true.[174] What as many experts as possible claim is considered truth.[175] Truth is therefore what the competent experts consider it to be. It is not an objective and immutable fact, but depends on the historical development of intersubjective knowledge. Representatives of the constructivist theory of science of the 20th century define it in no different way.

How does one gain the knowledge of the experts? Since it is said in the *Criticón* that wisdom has long fled from this world, as it no longer had any happiness or prestige among people and traces of it could only be sought and found in old books[176], reading is recommended on the one hand. On the other hand, social interaction is also an option. It is not books or school lessons that are helpful, but conversation with others, the "erudita conversación".[177]

Since one now arrives at truth not on the basis of one's own, but of others' experience, the problem of recognition becomes a problem of communication. Because according to Gracián important truths should not be communicated unencrypted. Only in indirect form are they taken seriously.[178] What is particularly easy to understand is given little importance. The incomprehensible is revered.[179] That Gracián adopts this standpoint as a writer is known to his readers from experience. He announces in his preface the *Discreto* as the art of the initiated. The mysteriousness of his style is meaningful, as it should not be understandable to all.[180]

So according to Gracián the paradoxical situation arises that knowledge is usually not given through one's own experience, but primarily through the oral and written experiences of others. Instead of these being accessible in a simple form, however, they are encrypted, so that only the initiated understand them. This means that truth eludes the access of communication no less than that of one's own experience. The question that arises here is whether

---

[173] Gracián 1986, 365.

[174] Gracián 1986, 392.

[175] Gracián 1986, 399; cf. Schröder 1966, 100.

[176] Gracián 1967, 738–739.

[177] Gracián 1986, 257–258.

[178] Gracián 1986, 270–271.

[179] Gracián 1986, 460.

[180] Gracián 1986, 257–258.

Gracián laments these difficulties of finding knowledge or whether he merely exaggerates them rhetorically to make them more interesting and important.

After looking at the difficulties that Gracián sees as fundamental to any knowledge, let us now consider the three central objects of knowledge: the self, the world, and others. The questions, whose answers lead to self-knowledge, are posed by Andrenio at the beginning of the *Criticón*, when he asks about his existence, his origin, and the meaning of his life:

> Pero llegando a cierto término de creer y de vivir, me sateó de repente un tan extraordinario ímpetu de conocimiento, [...] haciendo una y otra reflexión sobre mi propio ser. ¿Qué es esto?—decía. ¿Soy o no soy? Pero, pues vivo, pues conozco y advierto, ser tengo. Mas si estoy, ¿quién soy yo? ¿Quién me ha dado este ser para qué me lo ha dado? ¡Para estar aquí metido, grande infelicidad sería! ¿Soy bruto como éstos? Pero no, que observo entre ellos y entre mí palpables diferencias.[181]

Gracián's questioning and doubting self thus comes to a similar self-assurance as the doubting self of Descartes, in which it is said: I think, therefore I am.[182] "Pero, pues vivo, pues conozco y advierto, ser tengo" is stated by Gracián, who in turn derives the being of the self from thinking. One could translate Gracián's sentence as: I live, think, and perceive, therefore I am. However, Gracián does not stop at the methodical doubt of Descartes, who immediately leaves the position of doubt and builds a reliable world system with the "res cogitans" and the "res extensa". Gracián's doubt, on the other hand, is never completely lifted and appears in various places. Thus, Andrenio's important questions seem to be posed, but not answered.

Descartes believed that the world can only be recognized on the basis of self-knowledge. Gracián also requires self-knowledge to precede the knowledge of things.[183] Whether one knows something or knows nothing and must listen to others who know, at least that one should know about one's own self.[184] It can be concluded that Gracián fundamentally adopts a skeptical attitude towards the possibility of recognizing the self and the world. In the *Criticón*, Gracián emphasizes that the dignity of man lies in recognizing his own limitations and at least overcoming the state of barbarism with the help of the humanities, especially since one cannot strive for absolute knowledge.[185] Self-knowledge is also a prerequisite for dealing properly with the passions that hinder knowledge.[186]

---

[181] Gracián 1967, 523–524.

[182] Descartes 1971, 26.

[183] "Y quien comienza ignorándose, mal podrá conocer las demás cosas." Gracián 1967, 597.

[184] "Saber o escuchar a quien sabe. Sin entendimiento no se puede vivir, o proprio o prestado; pero hay muchos que ignoran que no saben y otros que piensan que saben no sabiendo." Gracián 1986, 430.

[185] Egido 2001, 21.

[186] Gracián 1967, 599.

In addition, self-knowledge is a precondition for the purification of the self from false notions. Here too, Gracián moves in lines of thought that are comparable to those of Descartes. While Descartes excludes everything that does not withstand his critical doubt as certain, in order to build an even more secure edifice afterwards[187], Gracián refers to the ancient philosopher Antisthenes, in order to strip off harmful things and learn correct ones after his model.[188] First and foremost, therefore, must be the teaching of the debunking of illusions and the exclusion of non-knowledge.[189]

If what constitutes the entirety of the self remains unrecognizable, Gracián at least considers those elements of the self that influence the understanding of the object world or other subjects. The fact that it is the subject that structures the object world in the Kantian sense becomes clear in Gracián when he attributes colors not to the objects, but to the subjective ways of perception, as the *Veedor* shows.[190]

So it depends on the subject's state of mind and not on the effects of the object when deciding what color something has. With this, Gracián emphasizes the unreliability of sensory perception and again represents a very subjectivist standpoint, as color perceptions do not follow the effects, but the affections.

Influences of the subject are not only evident at the level of sensory perception, but also at the level of judgment formation. There are virtues, the presence of which promotes knowledge and the absence of which prevents knowledge. For example, through lack of moderation, the greatest right becomes an injustice.[191] One can also think of the development of personality from youth to old age. It appears characterized by the increase in understanding, as well as by the purification of the will. However, Gracián points out that some never fully complete their development[192]—which again feeds doubt and skepticism regarding the possibility of truth recognition.

There is also a difference in the development of the knowing and the ignorant: While the *discreto* selectively takes nectar from the most beautiful flowers on his travels through the world, the ignorant remains ignorant due to lack of understanding.[193] But whether the ignorant becomes the knowing often does not depend on him. Whether a king is wise and capable, for example, Gracián makes dependent on the peoples he rules. While effeminate customs

---

[187] Descartes 1960, 19–20.

[188] "Encargaba, pues, Antístenes a sus tirones desaprender siniestros para mejor, después, aprender aciertos." Gracián 1986, 325.

[189] "¡Oh gran maestro aquél que comenzaba a enseñar desengañando! Su primera lición era de ignorar, que no importa menos que el saber." Gracián 1986, 325.

[190] Gracián 1967, 904.

[191] Gracián 1986, 393.

[192] Gracián 1986, 312.

[193] Gracián 1986, 375.

prevailed among the Assyrians, the Spartans set an example of efficiency for their kings.[194] But for all, they are born as raw, uneducated people and only develop into personalities through knowledge.[195] The wealth of knowledge stored in each individual subject is a good starting point for future insights— a kind of hermeneutic pre-understanding[196] Gracián grants to the educated, whose knowledge helps in the assessment of words and deeds.[197] Thus, it becomes apparent that according to Gracián, every insight depends on the prerequisites of the subject. Regardless of whether these have a beneficial or detrimental effect, an objective and universally valid insight appears to be excluded according to Gracián.

Gracián proves that the understanding of the world is not possible and has not yet been realized by inference. Complete knowledge of the world would indeed allow us to create it in its entirety, or at least parts of it. Even if we were to combine all future and past understanding, this would not be possible. Therefore, concludes Gracián, the understanding of the world is not given.[198]

If knowledge does not even seem sufficient for the hypothetical creation of the smallest part of the world, it should at least be possible to use existing knowledge to make improvements. However, Critilo is rather skeptical about the possibility of changing the world through the understanding. [199] Wisdom, therefore, ultimately consists in the insight that the world is fundamentally unknowable, since the criterion for knowing is the ability to recreate, which leads the wise Critilo to the resigned motto "see, hear, and be silent" (»Ver, oír y callar«).[200] Knowledge acquired through seeing and hearing should therefore not be shared, but kept silent. Here, Gracián's skepticism extends to the interaction of subjects with each other, which will be discussed in the next section.

In communication, where the goal is the transmission of knowledge, Gracián recommends avoiding clear statements and prefers obscurity and encryption. The same applies to the subject's interaction with other subjects

---

[194] Gracián 1986, 182.

[195] Gracián 1986, 395.

[196] Gadamer 1972, 252–256.

[197] Gracián 1986, 259.

[198] "¡Qué mucho—dijo Critilo –, pues, si aunque todos los entendimientos de los hombres que ha habido ni habrá, se juntaran antes de trazar esta gran máquina del mundo y se les consultara cómo había de ser, jamás pudieran atinar a disponerla! ¿Qué digo el universo? La más mínima flor, un mosquito, no supieran formarlo." Gracián 1967, 525.

[199] "El mismo fué [el mundo] siempre que es. Así le hallaron todos y así le dejaron [...]". Andrenio: "Pues, ¿cómo hacen para poder vivir, siendo tan cuerdos?" Critilo: "¿Cómo? Ver, oir y callar." Andrenio: "Y no diría desa suerte, sino ver, oír y reventar." Critilo: "No dijera más Heráclito." Andrenio: "Ahora dime, ¿nunca se ha tratado de adobar el mundo?" Critilo: "Sí. Cada día lo tratan los necios." Andrenio: "¿Por qué necios?" Critilo: "Porque es tan imposible como concertar Castilla y descomponer a Aragón.« Gracián 1967, 573.

[200] Gracián 1967, 573.

in general. Here, Gracián repeatedly advises against the clear communication of truth. In the encryption strategies that Gracián recommends, one can recognize not only elements attributed to Tacitus, but also the means of secrecy used by Botero in the interest of raison d'état.[201] With Gracián, it is about power at court, which is shown in the hierarchical relationship between superiors and subordinates.[202] This results in words having no meaning independent of the situation. Semantics proves to be rooted in pragmatics. The art of pleasing makes it possible for the same statement to be understood as flattery by one person and as an insult by another.[203] The speaking situation can even result in everything actually having the opposite meaning, namely when some tacticians speak negatively about something in front of others because they want it themselves. In this case, yes means no and no means yes.[204] The pragmatic assessment of advantage determines behavior, not orientation towards truth. This is particularly evident when Gracián advises against impressing the ignorant with the truth, as they would feel intimidated and react aggressively.[205]

In this case, it is better to withhold the truth. In general, according to Gracián, it is better not to reveal oneself when dealing with others. This makes it easier to see through the other person.[206] Even if special praise would be appropriate, social caution dictates that this should be withheld, otherwise one would not be taken seriously.[207] This can be referred to as the rhetoric of silence, which is demonstrated in Gracián's *Criticón* in its various forms.[208]

If one does not want to reveal the truth, one can therefore withhold it without lying. Withholding it requires just as much skill as telling it.[209] No less important than the art of concealing something, according to Gracián, is the art of concealing the fact that one is concealing something.[210] The principal difficulties of knowledge, such as the diversity and changeability of things, which were identified as inhibiting knowledge, are consciously used as means

---

[201] Lasinger 2000, 209.

[202] Giammusso 1993, 307.

[203] Gracián 1986, 335.

[204] Gracián 1986, 459.

[205] "Todo vencimiento es odioso, y del dueño, o necio o fatal. Siempre la superioridad fue aborrecida, cuanto más de la misma superioridad.« Gracián 1986, 361.

[206] "llevar sus cosas con suspensión. [...] El jugar a juego descubierto ni es de utilidad, ni de gusto.« Gracián 1986, 359.

[207] "El que alaba sobrado, o se burla de sí o de los otros.« Gracián 1986, 123.

[208] Egido 1991, 21.

[209] "Sin mentir, no decir todas las verdades. [...] Tanto es menester para saberla decir como para saberla callar.« Gracián 1986, 432.

[210] "Afectó Tiberio el disimular, pero no supo disimular el disimular. Consiste el mayor primor de un arte en desmentirlo, y el mayor artificio, en encubrirle con otro mayor.« Gracián 1986, 149.

of concealment in interpersonal interaction. Thus, one should not offer the other person a target through constant and predictable behavior.[211]

The primary concern is not the transmission of true information, but the creation of a specific conversational situation and a state of mind of the listener. The truth is subordinate to pragmatics and rhetoric. The other person's expectations should be maintained through constant variation, so that they do not turn away out of boredom.[212] The interaction of people with each other, according to Gracián, is most similar to a battle, for example the conquest of a castle, whose gates are to be spied and stormed by the attackers and hidden and defended by the defenders. The defender must not reveal anything of what the attacker wants to conquer. Therefore, the individual must conceal his affections, as they can be used to take over the rest.[213] Communicative interaction therefore primarily serves not the transmission of truth, but the tactical and strategic maintenance or acquisition of positions in a battle of all against all. It is a kind of martial art against the wickedness of people, which dictates never to reveal intentions.[214] This means that communication primarily does not serve the transmission of truth. The question arises whether this applies only to the interaction among courtiers or also to the situation of education, which Critilo and Andrenio exemplify in the *Criticón*.

Since recognition is primarily mediated through communication, it is important to know whether the specific form of communication, which represents the student's instruction by a teacher, promises success. Are the educational attempts that Critilo allows Andrenio to experience portrayed as successful by Gracián? For Gracián, communication is the prerequisite for the transmission of knowledge. The wise one imparts his wisdom through language.[215] Therefore, the first thing Critilo must do is teach Andrenio the language, which is easily achieved.[216]

In the numerous subsequent situations, however, it becomes apparent that despite mastering the same language, no real communication arises between Critilo and Andrenio. When Andrenio elaborates on his past experiences and insights, Critilo repeatedly interrupts him to comment on the meaning and

---

[211] "No siempre de primera intención, que le cogerán la uniformidad, previniéndole, y aun frustrándole, las acciones. Fácil es de matar al vuelo el ave que le tiene seguido, no así la que le tuerce.« Gracián 1986, 366.

[212] "Saberse atemperar […]. Siempre ha de haber novedad con que lucir, que quien cada día descubre más, mantiene siempre la expectación y nunca llegan a descubrirle los términos de su gran caudal.« Gracián 1986, 383.

[213] "Sabido los afectos, son sabidas las entradas y salidas una voluntad, con señorío en ella a todas horas.« Gracián 1986, 113.

[214] "Milicia es la vida del hombre contra la malicia del hombre. Pelea la sagacidad con estratagemas de intención: nunca obra lo que indica, apunta, sí, para deslumbrar.« Gracián 1986, 363–364.

[215] Gracián 1967, 522.

[216] Gracián 1967, 522.

deeper significance in each case. However, Andrenio never responds and continues his monologue unilaterally, as if his communication partner Critilo had said nothing. This is particularly noticeable when Critilo, on the occasion of Andrenio's discovery of the moon, evokes its function as a timekeeper and as a symbol of human imperfection, when Critilo praises divine providence after Andrenio's report on the distribution of the ripening times of fruits over several months; or when Critilo interprets Andrenio's remarks about his joy at blooming and sorrow at wilting flowers allegorically as an image of human transience. Andrenio always continues as if no one had spoken to him.[217]

Perhaps in the specific pedagogical situation, in addition to theoretical instruction, practical experience is required? Can Critilo's educational attempts succeed solely by conveying mere instructions to Andrenio? Perhaps what is said is understood, but it is not sufficient for complete knowledge, for which the recognizer's own experience would be necessary. Critilo thus says that human malice can only be understood through personal experience.[218]

Instruction also takes place in a playful manner. When Andrenio learns that at the court of truth not even the smallest lie or the mere withholding of truth is possible, he refuses to enter it.[219] Normally, however, everyone wants to be told the truth, without having to do so themselves.[220] And Critilo believes that the court of truth is not of this world.[221] If initially there is agreement to avoid the court of truth, as it contradicts widespread habits, Critilo corrects this with the hint that such a court can only exist in the hereafter. While the difference between immanence and transcendence is highlighted in a playful way, Critilo uses the surprising leap from theater fiction to the real world in another case: Along with a large crowd, Andrenio enjoys a play that depicts the fate of a man who is gullible and deceived by all. Critilo then suggests that Andrenio should rather be sad, as he himself, like the man portrayed, falls victim to all worldly deceptions, when the world deceives him, life lies to him, fate mocks him, health fails, and age passes.[222]

At first glance, *Criticón* gives the impression that everything is learnable for humans, that they should therefore learn everything and can develop into the ideal of a personality that corresponds to the humanistic educational ideal. However, upon closer inspection, clear limitations emerge. Firstly, in *Criticón* too, the physical disposition proves to be insufficient for the perceptions of reality, especially since Critilo and Andrenio encounter beings who have better developed senses, such as Argus with his many eyes.[223] This applies even though Critilo and Andrenio acquire numerous perspectives during their

[217] Gracián 1967, 530–534.
[218] Gracián 1967, 542/2.
[219] Gracián 1967, 873.
[220] Gracián 1967, 870.
[221] Gracián 1967, 876.
[222] Gracián 1967, 586.
[223] Gracián 1967, 671.

journey.[224] Moreover, the educator Critilo also cannot explain many things he sees and needs the support of outsiders for explanation. This is the case, for example, when they see bound and robbed people, but no robbers[225], when they have the panoramic view of the world from Argos explained to them[226] or even when Critilo is temporarily deceived by Circe's trick.[227] Finally, significant progress towards the perfect personality and the unmasking of deceptions can only be made shortly before death, so that humans no longer benefit from it.[228] Thus, the fact that knowledge, if ever achieved, is immediately lost again, reinforces doubts about its meaning.

Critilo is indeed trying to instruct Andrenio, but the great mass is consciously abandoned and left in their ignorance. For example, at a crossroads where the middle path is to be chosen and the extreme paths avoided, Critilo only pays attention to Andrenio. He doesn't care that the great mass takes the wrong path.[229] When Andrenio once wants to help pitiful slaves, Critilo stops him with the hint that they are the powerful, slaves of their desires and pleasures. This is enough to make an attempt at liberation seem unnecessary.[230] Once, Critilo warns Andrenio of a poisoned spring that destroys the good character of man, but this does not prevent him from calmly watching other travelers drink from it and be lost forever.[231]

The example of education through communication, as demonstrated by Critilo and Andrenio, thus does not appear to be successful: Despite having learned to master language, Andrenio's ability to understand is poorly developed, especially since theoretical instruction without practical experience remains empty. Even playful variants of instruction, such as the Court of Truth or the play, only show Andrenio's difficulties in distinguishing between immanence and transcendence, between lie, fiction, and reality. Not even Critilo proves to be a reliable educator when he is inferior to others in terms of his sensory perceptions and his judgment. Fundamental skepticism arises from the perspective that successful education ends in its loss through death. Educational successes are therefore limited and finite, the perfect transmission of truth only exists in transcendence. If these limitations apply to the privileged relationship between Critilo and Andrenio, then a relationship between Critilo and the great mass is simply avoided, so that not even an attempt at instruction is made here.

Thus, the ignorant are those whom Critilo does not care about. They form a large group in Gracián: "cualquiera necedad es vulgaridad y el vulgo

---

[224] Gracián 1967, 679.

[225] Gracián 1967, 613.

[226] Gracián 1967, 882; see also 683.

[227] Gracián 1967, 638.

[228] Gracián 1967, 977.

[229] Gracián 1967, 555–559.

[230] Gracián 1967, 567.

[231] Gracián 1967, 576–578.

se compone de necios."[232] The ignorant do not think about their goals or interests, lack diligence, spend a lot of time on the unimportant and little on the essential.[233]

In fact, the wise and the ignorant crowd stand opposed as representatives of two irreconcilable parties. This is evident in thoughts like: If something pleases everyone, it cannot be perfect, as perfection is the matter of only a few people.[234] The circle of those belonging to the ignorant crowd seems to be very large.[235] However, all parties seem to be so convinced of themselves and their opinions that they have no arguments for outsiders, but only ridicule.[236] Gracián allows the wise to enjoy the folly of others.[237] While the wise in principle discovers, notices, achieves, understands, and defines everything by his nature, it is quite different with the large crowd: "raras veces discierne entre lo aparente y lo verdadero; es muy común la ignorancia y el error muy plebeyo."[238]

Does the derogatory treatment of the large crowd result in the unrestricted praise of the wise? That even the wise are not really knowledgeable, Critilo and Andrenio have to learn in Rome on the occasion of a disputation about human happiness. The wise speakers all come to different conclusions and can convince the audience only until the next one starts to speak. In the end, they have to be taught by a madman that they make the mistake of seeking on earth what can only be found in the hereafter.[239] Here the old Christian dialectic comes into play, that the one who appears mad in this world is the truly wise in the transcendent background.

It remains to be noted that the wise is not competent everywhere. This is especially true in everyday practice, where he does not see through the deceptions of others. While he knows the extraordinary, he does not know the ordinary.[240]

---

[232] Gracián 1986, 442.

[233] "No pensando, se pierden todos los necios: nunca conciben en las cosas la metad, y como no perciben el daño, o la conveniencia, tampoco aplican la diligencia. Hacen algunos muchos caso de lo que importa poco, y poco de lo que mucho, ponderando siempre al revés.« Gracián 1986, 373.

[234] "Pésele de que sus cosas agraden a todos, que es el señal de no ser buenas: que es de pocos lo perfecto.« Gracián 1986, 458.

[235] "Son tontos todos, los que lo parecen y la metad de los que no lo parecen. [...] pero el mayor necio es el que no se lo piensa y a todos los otros define.« Gracián 1986, 440.

[236] "Tienen muchos por felicidad gozar de lo que apetecen, condenando a infelices los demás, pero desquítanse estos por los mismos filos; con que es de ver la mitad del mundo riendose de la otra, con más o menos de necedad.« Gracián 1986, 122.

[237] "Mire desde la talanquera de su cordura los toros de la necedad ajena.« Gracián 1986, 300.

[238] Gracián 1986, 320–321.

[239] Gracián 1967, 958–960.

[240] "Tener un punto de negociante. No todo sea especulación, haya también acción. Los muy sabios son fáciles de engañar; aunque saben lo extraordinario, ignoran lo ordinario del vivir, que es más preciso." Gracián 1986, 452.

Since the knowledge of the wise does not enjoy high recognition[241], the wise will not befriend the ignorant.[242] Nevertheless, the wise should bear the ignorant not with impatience, but with calmness.[243] If the wise wants to live in peace with the ignorant, he may pretend to be ignorant himself.[244] To avoid being misunderstood, Gracián explains, the wise should only pretend to be ignorant, without actually becoming so.[245] After all, the wise can only be heard as ignorant among the ignorant.[246] The gap between the knowledgeable and the ignorant thus seems insurmountable. And it seems rather unlikely that, with the recommended distancing and pretense, the wisdom of the wise will become recognizable or apparent. It does not seem that Gracián follows Plato's allegory of the cave and trusts the wise to convey his wisdom to the great multitude.

Consequently, the advice is that the wise should keep his secrets deep within himself.[247] Seneca has already developed the stoic concept of the autarky of the wise, who rests within himself like Jupiter. Since he carries everything he needs with him, it is pointless for him to go public, for no one would understand him.[248] However, when a wise man encounters a second, the situation looks different. Then a rather counterfactual ideal case occurs, which should be characterized by a gem. For the wise only associate with the wise, because the truth must not be entrusted to either malice or ignorance.[249] So if the wise are to keep to themselves and the ignorant are to do the same, then the latter will always form the majority and the one will despise and mock the other. If the wise is to amuse himself over ignorance, but only show himself ignorant among the ignorant and keep his knowledge to himself, then the truth will never be communicated to the ignorant. This is not changed by the weaknesses of the wise, which are not comparable to those of the ignorant.

The contrast between the wise and the ignorant multitude, like the relationship between Critilo and Andrenio, has shown that the transmission of knowledge through informing or educating is difficult or impossible. The view of communicative interaction using the example of courtiers showed that truth is displaced by advantage, by tactics, silence, and stratagem. Semantics is dominated by pragmatics. The recognition of the world appeared insufficient, as it was neither convertible into practical

---

[241] Gracián 1986, 399.

[242] Gracián 1986, 412.

[243] Gracián 1986, 423.

[244] Gracián 1986, 412.

[245] Gracián 1986, 455–456.

[246] Gracián 1986, 377.

[247] Gracián 1986, 431.

[248] Schröder 1966, 88.

[249] Gracián 1967, 323.

reconstruction nor useful for changes. The self, which derives its existence and its identity, which distinguishes it from animals, from its recognition and perception, which is supposed to cleanse itself of false ideas in order to be able to take up the true ones, whose subjective perceptions, virtues, stages of development, and prior knowledge constitute the object world, is ultimately not recognizable and seems to influence the recognition of objects more than the object itself, leading to subjectivism and relativism.

It has become clear that the principal difficulties of knowledge demonstrated at the outset are often based on the tropes of Diogenes Laertius. Since knowledge is complicated by the diversity of things and their aspects, since the substance cannot be deduced from the accidents and the number of ciphers is infinite, only the imagination can construct images and signs of them independently from the thing in itself. However, the subject is no less diverse if the perception of one differs from that of another, if the same sensory perceptions are interpreted differently, and if, in the face of general relativism, which makes truth intersubjectively dependent on positions, as well as in the face of the constant developments and changes of the subject, uncertainty and flexibility are the consequences. It has also been shown that cognition is not independent: It is promoted and endangered by the will, by interests, by utility, passion, and preferences. Gracián generalizes the principle of orienting oneself in uncertainty according to the opinion of the supposedly knowledgeable and gives special importance to knowledge through the mediation of experts. If one now arrives at truth through reading and interaction with others, then the problem of knowledge becomes a problem of communication, which Gracián, however, does not want to facilitate but to complicate, otherwise the communicated would be considered worthless. Thus, it becomes apparent that Gracián's skepticism concerns not only the knowledge of one's own subject but also that of the world of the object and other subjects, but it particularly starts with communicability. And this is all the more serious as, according to Gracián, knowledge primarily takes place through communication.

With Francisco Sánchez and Montaigne, it was primarily the sensory perceptions that lead to false knowledge, which becomes clear when one considers that some animals have better sensory organs, or imagines what human knowledge would look like with fewer sensory organs. Gracián agrees with this criticism. Just as Descartes prefers his own introspection to the ancient authorities and excludes everything that does not withstand his critical doubt, so Gracián also wants to strip off apparent truths. The multiplicity and changeability of things is opposed by the changeability of the recognizing subject, which results not only from the age of life but also from the respective interest and the guiding affects.

If truth is now what competent experts consider it to be, then this must be communicated to others. But if this now depends on the partners involved, the power relations, rivalries, and struggles prevailing among them, then the situation, according to Gracián, can be illustrated with the image of the conquest of a castle, where one has to defend oneself against one's attackers. The wise cannot simply instruct the ignorant, as he is not recognized by them. According to Gracián, he protects himself best by keeping a low profile, which in the end rather prevents the truth from spreading.

2.5   PSEUDO-KNOWLEDGE OF THE IGNORANT

While skeptics doubt the possibility of certain knowledge, there are others who are convinced of their knowledge. They do not consider that it could be shaped from a subjective perspective and that different views are possible. From the perspective of the wise, the knowledge of the specialist appears all too fragmentary. Knowledge can be discredited if it was gained with inadequate methods, if the participants lack ethical standards and are guided by fame, vanity, and profit, or if its application is accompanied by negative consequences. Fools have unimportant or ridiculous knowledge.

If on the one hand the fool is the opposite of the wise, on the other hand the madman differs from the simple fool in that others negate his normality. The actual or apparent advantages of madness are praised by Erasmus of Rotterdam in his *Encomium moriae* (1510). Whole pages of the Erasmian original can be found in the Spanish adaptation by Jerónimo de Mondragón in his *Censura de la locura humana y escelencias della* (1598), a work that, like many others, shows the influence of Erasmus on Spanish literature. Mondragón presents two types of madmen: Some are the focus of the masses' ridicule, are caricatured and considered mad. They are the ones he wants to praise because they seem particularly wise and clever to him. In contrast are those who the common people consider wise and rational, even if they are actually mad. As an example of a wise man whom the foolish crowd considers mad, Mondragón tells the story of the inhabitants of Abdera, who asked Hippocrates to cure Democritus, who suffered from incessant fits of laughter, of his madness. When asked the reason for his suffering, he replied that he could not help but constantly laugh at people because he considered them mad. Hippocrates was convinced and told the inhabitants of Abdera that they were the ones who needed to be cured of their madness and gave them the herbs he had originally collected for Democritus. Mondragón takes another example from the Bible. According to the Bible, those who are considered wise by the majority are in reality the mad ones. Most consider them wise, even though they are morally corrupt and therefore insane.

According to Mondragón, the foolish majority only chooses the fool, i.e., the false wise. Thus, those who are considered mad by the majority are in reality particularly wise because they are not constantly occupied with eating, drinking, clothing, honor, fame, and pomp. But these are the occupations of those whom the people consider wise and who live with such restlessness that they should actually be considered mad. This group also includes schol-ars who let themselves be carried away by their desire for fame and immerse themselves in the arts and sciences with sweat and tears night after night.[250] Mondragón uses ethical criteria when he intends to criticize the excessive pur-suit of fame and the arrogance of those who are considered scholars, among whom are certainly many of his humanist colleagues. On the other hand, he also wants to show that in general it is not those who are wise who are rec-ognized as wise, but those who are considered such by the great majority. However, the majority cannot decide who is wise and who is not. Therefore, the majority judges completely wrongly and often gives the wrong ones the honor of wisdom. The target of Mondragón's criticism here is the majority, which lacks the necessary judgment.

Like his predecessors, the diplomat and writer Saavedra Fajardo also crit-icizes the disciplines and their representatives. Where in his first version of the work *República literaria* (1613–1620) he lets the figures Democritus and Heraclitus speak, he adopts on the one hand the satirical register of Mondragón and on the other hand the argumentation of Erasmus and Vives. Democritus can only react by laughing at the misdevelopments and inade-quacies, as they are so numerous that he cannot change them, starting with the useless activities that are so dear to the humanist circles. Thus, he mocks the figure of the humanist who spends his years in vain trying to waste his mind with the reading of medals, the study of ancient stones, the inspection of ruins and building parts, and trying to find out which shoes Cadmus wore using various manuscripts.[251] According to him, the fact that a humanist, for example, invents puzzles after the humanistic interpretation of hieroglyphs or deals with anagrams, translations, or glossaries brings neither honor nor profit, but only fatigue. For Democritus, these occupations are ridiculous, as is the compilation of sentences by ancient authors so loved by the human-ists; it is as if one were removing stones from a building, which, taken out of context, lose their meaning. This would only promote laziness and save the reading of the entire work. Also mocked are authors who dedicate their works to kings and the powerful, as if they could approve something through their brilliance that they do not even understand. So too are those who are convinced that they are considered learned because they own a large library or grow a beard like an ancient philosopher. Criteria for judgment are thus the relevance of scientific activities and the ethics of scientists.

Saavedra Fajardo assigns Democritus the task of examining the liberal arts one by one. Grammarians are distinguished by their vanity. Just because they know something about grammatical genres and verb forms, they consider

---

[250] Mondragón 1953, 49–53.
[251] Saavedra Fajardo 2006, 152.

themselves competent in other sciences and capable of criticizing, by calling Plato confused and Aristotle obscure, and criticizing Pliny's accumulation of details without sense and order. Rhetoric is accused of a self-loving attitude and the effect of flattery or tyranny, as well as the art of captivating and lying, by achieving with gentle force what is not possible for the truth.[252] Since they can deceive and dazzle the people, the representatives of rhetoric are banished in both Plato's Republic and ancient Rome. Poetry, the sister of rhetoric, was expelled in the same way because it puffs itself up against the other disciplines and claims supremacy; in reality, it only lies, drives away the truth, limits itself to entertainment, and is useless.

Instead of continuing with dialectic, the third discipline of the trivium, Democritus next turns to history and moral philosophy, two disciplines that were added to the trivium in the early modern period. Historians would report more about the vices of kings and the powerful than about their virtues, simply because there are more vices in the world. In view of this, the reader prefers to consider the vices of others and thus excuse his own, rather than imitate the virtues, if there are any. For Democritus, it provokes particular laughter when historians presume to be competent in political theory just because they know a few numbers and details. Moreover, historians cannot be reliable in their reports because they have to invent the causes of historical developments or resort to general opinions, as they cannot be present at the secret meetings of princes where the motives for decisions are discussed. Furthermore, according to Democritus, since they are not nobles, they cannot put themselves in the position of the princes and cannot recognize their true motives, virtues, and vices with their limited imagination. Moral philosophy, on the other hand, provokes laughter solely because of the multitude of its schools, which pursue human happiness in different ways: Epicurus through pleasure, Aristotle through virtue, Theophrastus through bravery, the Peripatetics through speculation, Periander through fame, power, and wealth, and the Platonists through the highest good.

The disciplines of the quadrivium are also accused of pride, especially arithmetic, whose numbers encompass all other disciplines. And geometry is no less arrogant, even if its principles can more easily find general agreement. If astronomy does not find agreement, it is clear that on this basis the audacity of astrology can only provoke laughter when it dares to predict the future course of a person's life as inevitable, even though it is impossible for human ingenuity to know the properties of the stars.[253] Even less sense is made by

---

[252] Saavedra Fajardo 2006, 155.
[253] Saavedra Fajardo 2006, 162.

the disciplines that, like astrology, predict the future: chiromancy through the folds on the hands, geomancy through points, pyromancy through fire, and hydromancy through water.[254]

Democritus finds the higher faculties of the university no less ridiculous. This is also the case with the jurists, who compensate for their lack of authority by deriving their laws from divine law or natural law, or by demanding money for everything they say or do not say. The representatives of the medical faculty are no less interested in profit when they have to watch over the health of the powerful. Medicine as an empirical discipline is viewed very critically because its knowledge is provisional and incomplete and suffers not least from the possible deceptions of the senses. The inaccuracy of medical diagnosis shows how little medicine or those who profess it can achieve.[255]

Theology, the third of the higher university faculties, is not discussed due to prevailing censorship. For this reason, Democritus makes some final considerations. He does not mock the sciences as such, but their own representatives in his republic, who are presumptuous and arrogant and are considered wise by the foolish majority. In reality, however, only the one who despises both the opinion of the majority and the riches and honors dependent on external circumstances, who focuses on his own reason and free will and controls his passions, is truly wise. Thus, the wise man is almost equal to the gods, while the foolish and arrogant, the representative of science, is nothing more than an animal, the wildest that the forests produce, for in inhumanity and wildness of his rebellious spirit, he is no different from it.[256] Heraclitus, who had listened to Democritus, completes the idea by pointing out the inadequacies of the scientific disciplines, which, however, saddens him more than it makes him laugh.

It can be determined that the satirical and critical representation of the disciplines and their actors takes place on various levels. Thus, the context of knowledge can be the criterion for whether knowledge is acceptable or not. As shown with Erasmus and Vives, the knowledge of antiquity can be relativized or denied in a Christian context; a phenomenon that, as we have seen, is criticized by Erasmus. Mondragón uses the Bible as a criterion to decide who is truly wise and who is truly mad. The different areas of the subjects of the disciplines can also be responsible for these disciplines leading neither to knowledge nor to understanding. The future of things can neither be read in the stars, nor in the lines of the hand, nor in the fire, nor in the points, nor in the water as emphasized by Vives and Saavedra Fajardo. The reason they reject the experiments of the disciplines in question in this sense is that

---

[255] Saavedra Fajardo 2006, 169.

[256] Saavedra Fajardo 2006, 169.

[254] Saavedra Fajardo 2006, 163.

they are unscientific. If a dubious goal is attributed to a science, that is reason enough to reject it as a whole. This is shown by Saavedra Fajardo when he attributes the goal of lying and deception to rhetoric and the goal of useless and deceitful entertainment to poetics.

The inadequacy of the methods can be blamed when the scientific character of a discipline is denied. Thus, Saavedra Fajardo uses numerous arguments when he denies the historian the ability to abstract in political theory and certifies him the inability to recognize the backgrounds and causes of historical events, which he cannot uncover because he was not there. Moral philosophy is presented by Saavedra Fajardo with such a number of proposed solutions that it is not possible to distinguish which one is the right one. Here, as well as in medicine, which as an empirical discipline runs the risk of being deceived by the senses, Saavedra Fajardo states false knowledge. He criticizes the jurists and accuses them of only referring to divine and natural law to gain respect and recognition. Closely linked to the method is the practical exercise of a knowledge discipline. Thus, philology can devote itself to important topics and texts, or it can waste its time with less important activities. The latter is what Saavedra Fajardo accuses the humanists of: compiling anthologies of sentences, seeking protection through dedications, or dealing with medals, ruins, riddles, glossaries, or unimportant individual questions.

Finally, the representatives of a discipline can be the criterion for determining its reputation. Erasmus noted with regard to Socrates that modesty increases proportionally with knowledge. Saavedra Fajardo cites as an example grammarians, whose knowledge is limited to tenses and grammatical genres, but who believe they are competent in all disciplines. He mentions arithmeticians, who see all disciplines contained in their numbers. Arrogance and presumption is what Saavedra Fajardo's Democritus had accused all representatives of the disciplines of. Precisely for this reason, they are considered wise by the ignorant majority, as it cannot form a judgment. The delicate relationship between the wise and the others was often mentioned; Vives advises the wise to stay away from the majority, as it only spreads false opinions; Mondragón distinguishes the fools, who are considered wise by the masses but are in reality mad, from those who are considered mad by the majority but are in reality wise. As an example, Mondragón cites Democritus, who was considered a madman, but was in reality a wise man. Against this background, one can understand the constructions of a literary *res publica*, a society of true scholars, to which ideally only the wise belong. The Socratic union of knowledge and morality is finally postulated by Vives when he demands clean work from scholars, which is shaped by reason and not disturbed by passions such as anger, greed, and ambition. For Saavedra Fajardo too, the wise man is so alien to external conditions, wealth, honors, and passions that he is guided only by his own reason and free will. The wise man is therefore not contrasted here with the ignorant in order to mark an insurmountable barrier in a skeptical attitude. It is the knowledgeable, whether in less useful special fields, whether driven by passions, whether blinded by self-conceit and arrogance, whose knowledge is to be relativized in order to lose the appearance of validity and relevance through the standards of the wise.

CHAPTER 3

Human

3.1 CROWN OF NATURE

If knowledge is to be oriented towards humans, this is not least due to the outstanding position in the universe attributed to them in the early modern period, even when their greatness was compared with their insignificance. If now the *dignitas hominis* is opposed by a *miseria hominis*, then the latter indicates what needs to be overcome. The most important arguments for the precarious situation of humans had already been compiled by Pliny the Elder (23–79 AD), known for his encyclopedic *Naturalis historia*. Although nature had created everything else for his sake, man had to accept some disadvantages. He was not naturally protected against cold and heat, had to learn to speak and eat, was susceptible to diseases, and was the only being to suffer from fear, grief, superstition, debauchery, and ambition. Unlike animals, he usually experienced harm from his own kind.[1] The positive aspects of humans were addressed in the famous *Oratio de hominis dignitate*, published in 1496 by the Italian humanist Giovanni Pico della Mirandola. He considers humans a great wonder, to be ranked just below the angels. The fact that he is not fixed with his free will, but stands in the middle of the world, is to be seen as an opportunity. He can degenerate downwards into the animalistic, but he can also regenerate upwards into the divine of his own free will. Pico della Mirandola urges us to despise the earthly and to consider everything below the sky as insignificant, in order to devote oneself entirely to the spiritual and divine.[2] Empedocles is quoted, according to whom man has two natures, one of which lifts him up into the sky, while the other pushes him down into the underworld.

[1] Pliny 1975, 12–17.
[2] Pico della Mirandola 1997, 8, 15.

Starting from Pliny and Pico della Mirandola, in 1546 in Spain the humanist Pérez de Oliva composed a *Diálogo de la dignidad del hombre*, in which Aurelio represents the *miseria hominis* and Antonio the *dignitas hominis*. Since the controversy remains undecided, Francisco Cervantes publishes a sequel in favor of the *dignitas*.[3] Antonio Camos explains the *dignitas* by seeing the microcosm of man as hierarchically structured as the macrocosm of the universe and in his extensive work published in 1595 *Microcosmia, y govierno universal del hombre christiano, para todos los estados, y qualquiera de ellos*[4] he asserts perfection where the spirit rules. A preliminary remark about the meaning, possibilities, and benefits of instruction and instructive books is followed by a general consideration of the virtues of man and the respect he deserves.[5] Referring to *Genesis* of the Old Testament, Camos assumes a different degree of perfection for the individual types of creatures. The lower level is occupied by elements such as water, earth, or air. On this level, gold, silver, metals, and stones are also classified. Above this level are plants and trees, which are alive as they ripen, grow, and decay. A step higher are the animals, which not only exist and live but also have sensory perceptions. But at the highest level stands man as the most perfect being, who not only carries all the subordinate levels within him but also has reason, free will, and an immortal soul.[6] The world with its diversity serves man. As a result of his possession of intellect, he can invent sciences and gather experiences. Camos refers to Augustine when he sees man related to angels and God due to his intellect.[7] The kinship with God and the angels appears more brilliant and greater than that with the plants, animals, and all physical beings of the earth, thus devaluing the latter and thus the outer spatial world.

As an example, Gracián's *Criticón* is cited. There it says that all elements of the universe have contributed to the creation and perfection of man. Thus, the sky gave him the soul, the earth the body, the fire the warmth, the air the breath, the stars the eyes, and the sun the face.[8] Calderón also sees man as a microcosm composed of different parts of the universe. Thus, he has a connection to inanimate objects like stones, to plants that are born and grow, to birds that have living sensations, to angels that understand and speak, and finally to the divine eternity that characterizes his soul.

This assessment is to be seen in the early modern period against the backdrop of the narratives of the Bible. In *Genesis*, man is understood as the ruler over nature: "Let us make man in our image, after our likeness. They shall

---

[3] Strosetzki 1987, 16–18.

[4] Camos 1595.

[5] Camos 1595, 9.

[6] Camos 1595, 13.

[7] "Esse hombre es mas honrado, y de mas estima que el Angel, en razon del estrecho parentesco que quiso Dios contraher con naturaleza humana, suppo situandola con la persona diuina." Camos 1595, 16.

[8] Gracián 1967, 224.

rule over the fish of the sea and the birds of the sky, over the cattle and all the wild animals of the field and over all the creeping things that move on the earth. And God created man in his image."[9] Similarly, it is said in the *Psalms*: "You made him a little lower than the angels, you crowned him with glory and honor. You made him ruler over the works of your hands, you put everything under his feet."[10] When Adam was expelled from paradise, the situation did not fundamentally change. Although the fall from original perfection and height can be described as a decline, it can be seen as an opportunity from the perspective of a resurgence. The early Christian writer Tertullian compares the epochs of human history with the stages of human development from childhood to adulthood. The development thus optimistically goes from the less perfect to the more perfect.

In the hierarchy of the universe, man is thus the highest being, the center. Within him, he carries all other less perfect stages that occur in the universe. Only he has the most important thing: reason. His free will puts him in the position to become an animal or to overcome the disadvantages he has compared to animals and to become the ruler over the creation entrusted to him. It is obvious that with this focus, the universe is less interesting than man.

3.2   EXTERNAL WORLD AND INTERNAL WORLD

Does space belong to the recognizing subject or to the recognized object? This question is one of the oldest in philosophical epistemology. While the empiricists held the view that space and the objects found in it come into the human mind solely through perception and experience, the rationalists believed that space and the objects found in it are projected by the mind onto what we then call reality. Kant offered a mediating solution by considering space and time as what emanates from the human mind as a principle of order, but allowed perceptions and experiences to be influenced by external reality. So space is not just the external seen from the recognizing human being; the relationships and influences between subject and object are interpretable in many ways.

Paradigmatic for the 17th century is Descartes, who referred to space and the things in it as extended *(res extensa)* and contrasted the active mind as *res cogitans*, the latter of which is not spatial or extended. Descartes, as is well known, doubted everything in order to find a secure starting point, until he was left with only the certainty that he doubts, that there is therefore a *res cogitans*. Only starting from this certainty could he explain what a physical thing, a *res extensa* is: "By body I understand everything that is limited by any figure, that can be locally circumscribed and fills a space in such a way that it excludes any other body from it; what can be perceived by feeling,

---

[9] Genesis I, 1, 26–27.

[10] Psalm 8, 6–7.

sight, hearing, taste or smell, can also be moved in various ways, not by itself, but by something else that touches it."[11] *Res extensa* is also the human body, which in its extension is opposed to the soul, which in Descartes as *res cogitans* has no extension. But how does a non-extended soul relate to a spatially extended body, e.g., to its own? If the soul is understood as a body, albeit a particularly fine one, then its relationship to the human body is clear. The relationships are also clear if the body is understood as spirit or soul. The problem only arises when, with Descartes, one distinguishes between *res cogitans* and *res extensa*, assigning the former to the soul and the latter to the body. Then the question arises as to how something non-spatial can act on something spatial and how it is at all conceivable that such completely different things should be in a reciprocal relationship.

If man is considered the crown of nature, it is not surprising that his inner life is valued higher than the external world of objects surrounding him. This assessment began with Plato, for whom the external world of objects depends on a world of ideas. This view was adopted by Augustine and given a Christian context. It continues wherever the subject shapes the world of objects or at least is given preference.

Plato's ideas are independent and transcendental, thus distinguishable from the perceptible world. The allegory of the cave in Plato's *Republic*[12] describes the human journey through the world of phenomena and ideas in four stages of knowledge: words, perceptions, concepts, and ideas. Initially, he appears as if chained, looking at a wall. Only when he is freed from his chains and turns around, blinded by the fire, can he realize that the objects depicted there are only shadows of devices and figures moving behind him and illuminated by a fire. In the first phase, he only dealt with words, while in the second phase, he perceives empirical phenomena. In the third phase, he leaves the cave with its twilight and climbs up to the upper world with its daylight. He is no longer in the visible, but in the conceivable space and sees in the reflections of the stars on the earthly waters the concepts as images of the ideas. In the fourth and final phase, he turns his gaze away from the earth and directs it towards the sky, where he sees the ideas as stars and the all-illuminating sun as the highest idea. He realizes that it is the originator of everything he saw on his journey. Just as the sun enables seeing and being seen through its light, so the highest idea, the idea of the good and beautiful, is the reason for the recognition and being recognized of the ideas. In the dialogue *Parmenides*[13] this highest idea is the unity that unites all plurality in a higher sense within itself.

According to Plato's theory of ideas, all visible objects in the world depend on conceivable ideas, so that, for example, the particular chair on which one

---

[11] Descartes 1993, 23.

[12] VI 506 e 3 ff.

[13] 137 c 4–142 a 8.

is currently sitting owes its shape to the idea of a piece of furniture with an elevated seat and feet and thus participates in the idea of the chair. In the *Phaedo*[14] the idea is also referred to as the cause of the individual thing. There are two directions of consideration. The *Methexis* refers to the participation of the individual thing in the idea, while the *Parousia* means the shining through of the idea in the individual thing. According to Plato, the soul has seen everything in its true essence, i.e., as an idea, before birth. This it has forgotten after birth in the body, but can remember it again. Knowledge of the true essence is therefore not conveyed by sensory perception, but by the mind. If these ideas are now hierarchically ordered as in the allegory of the cave, a supreme idea appears from which all others are derived. For Plato, it is the idea of the good and beautiful. Everything that is beautiful is also good for him if it is derived from this idea. For a work of art, this means that its ethical quality corresponds with the aesthetic. An ethically bad work of art cannot be aesthetically beautiful. Just as the sun enables seeing and being seen through its light in the realm of the visible, so does the supreme idea in the realm of the conceivable.[15] Plato's theory of ideas has another consequence for art. If the concrete object is an image of an abstract idea, then this idea is more important than its concrete image, as numerous concrete objects orient themselves to it. But if the work of art imitates concrete objects, then it creates images of images of ideas, so it is even further removed from the idea than the concrete object and therefore less valuable than it. From this consideration, Plato derives his negative attitude towards that art in which, as in, for example, Homer's epics, the good is not depicted. Art is only justified if the idea of the good and beautiful shines through it.

In the early Middle Ages, with Augustine, the Platonic highest idea is equated with God, who created everything according to the ideas contained in him. Human knowledge, for him, is participation in the divine knowledge of ideas. By defending the monism of the good, like Plotinus, Augustine opposes the Manicheans, who see good and evil as equally original. When Augustine distinguishes the *civitas Dei* from the *terrena civitas, civitas* is to be understood as a community, not as a territory. The model is the Roman state, whose citizenship had a worldwide scope. One remained a citizen of Rome, no matter where one went. With the phrase "civis Romanus sum" the Roman identifies himself and associates with it the worldwide right not to be tortured or forced into labor. One should similarly imagine Augustine's *civitas Dei* and *terrena civitas*. He sees them as two communities that coexist from the beginning to the end of world history, with the former characterized by love of God to the point of despising one's own self, and the latter by self-love to the point of despising God.[16] Through his behavior, the individual

---

[14] 99 b 3 f.

[15] Republic VI, 506 e 3 ff.

[16] Redkova 2018, 158.

can thus assign himself to one or the other community. The material world, in the Augustinian sense, especially in cities, offers temptations and opportunities for pride and the pursuit of earthly power, which is why the patriarchs did not found cities, but lived as shepherds. Their *peregrinatio pro Christo* could then be modified by monks, like the Cistercian Bernard of Clairvaux, to *peregrinatio in stabilitate* and transferred to the monastic community, which was perceived as *civitas Dei peregrinans* because it offered a space free from the world with its material distractions, entirely serving spirituality.[17] Not without reason, Augustine advises a focus on the self, as that is where truth lies: "Noli foras ire, in teipsum redi; in interiore homine habitat veritas."[18] The path of the Platonic philosopher from the cave to the sun becomes the Christian *peregrinatio* with the Spanish humanist Fray Luis de León, whose path he describes. Here, it is about role model, imitation, and participation. Participation is achieved through imitation.[19] The path is also a moral law and commandment.

The Spanish *Siglo de Oro* is characterized by a devaluation, relativization, and subjectivization of the external world. If man as a microcosm contains the whole world within himself and derives his dignity from his likeness to God, if in the book of nature the will and understanding of the creator are primarily to be read, if the influence of the stars is limited to the body and the free will of man can overcome this, if the sensory perceptions of the eye can only perceive accidents and are therefore inferior to the sense of hearing, if not the eye, but only the mind and the soul can see, then space is largely a product of the recognizing subject.

Augustine assumes the self-assurance of the individual long before Descartes. By the soul knowing itself and about itself, it comes to the knowledge of God and what is real and valid. It is not the experience of the outer world of the cosmos that provides such knowledge. Therefore, for Augustine, the movements of the mind, conscience, and will are more significant than the relationships with the body and the outside world. In the face of a secondary outer world, all knowledge must originate from inner experience, which became largely prioritized in the Middle Ages. Franciscans like Roger Bacon emphasized the fundamental difference between spatial-material and psychic in the Augustinian sense, and highlighted that spiritual substances have no place in space. Instead, a tripartite division is made in humans between spatial-physical matter, body formed in the Aristotelian-Thomistic sense, and thirdly, the actual soul. The idea that one can come to the knowledge of God through introspection, through self-knowledge, was

---

[17] Redkova 2018, 158.

[18] Augustine 1962, 39, 72.

[19] "Cristo es el camino del cielo, porque si no es poniendo las pisadas en él y siguiendo su huella, ninguno va al cielo. Y no solo digo que habemos de poner los piés donde él puso los suyos, y que nuestras obras, que son nuestros pasos, han de seguir á las obras que él hizo." Luis de León 2008, 71–72.

later adopted by Descartes. He proceeds from self-consciousness and inner certainty to the idea of God and only then to the outside world. Finally, in Leibniz, introspection becomes the doctrine of monads, which as souls are completely self-contained and an inner world independent of external influence.

It is similar when mathematical or geometric formulas are applied to the *res extensa*. When man does this with his understanding, he imitates the creator God according to medieval understanding. The German natural philosopher and theologian Nicholas of Cusa writes in *De docta ignorantia I* that God used the same arts in creation that man also uses in his exploration of nature: "Est autem Deus arithmetica, geometrica atque musica simul et astronomia usus in mundi creatione."[20] If for Galileo Galilei the book of nature was written in mathematical language, Johannes Kepler said: "Ubi materia, ibi geometria."[21] If philosophical thinking was ultimately a kind of calculation according to the geometric method for Hobbes, the saying attributed to Einstein, "Not everything that can be counted counts, and not everything that counts can be counted," relativizes the importance of mathematics.

Since Euclid, geometry has to do with measures and proportions of things. In the Middle Ages, the natural philosopher and Aristotelian Robert Grosseteste analyzed the significance of light and color for the perception of objects within the framework of geometry in his work *De luce seu de inchoatione formarum*. Following him, Leonardo da Vinci, Johannes Kepler, and René Descartes later developed optics as a science within geometry. In the 14th century, Thomas Bradwardine in Oxford wrote a treatise on proportions in which he mathematically captured the speed of bodies, and in Paris it was Nicholas of Oresme who wanted to capture things graphically and quantitatively in the same century. This shows the scientific endeavor to describe the objects of nature mathematically, which found its particular expression in the Italian Renaissance with the polymath Leonardo da Vinci (1452–1519). He left behind hundreds of anatomical drawings and demanded physiological and anatomical studies as a prerequisite for the artistic capture of man. In doing so, he dissected and also studied pathological changes in the organs. In order to be able to draw, one should know the development of man from birth through growth to old age, his feelings from cheerfulness to sadness, and his various movements. In his studies, he combined experiment and mathematical approach as well as visual art and medicine. The following will therefore illustrate the mathematization of spatial experience using the example of visual art. Because sculptors and painters must first deal with spatial reality in a certain way before they can model it themselves.

Let's start by returning to Descartes. For him, it is not the eye, but the brain, that coordinates the visual impressions of both eyes into a single image.

[20] Eusterschulte 1997, 469.
[21] Sihvola 2000, 187–192.

He illustrated this in his work *La Dioptrique* with a blind man's stick, which, like the eye, is nothing more than a prosthesis of perception.[22] Just as the beam of light gives signals to the eyes, the blind man's stick conveys sensations, which have nothing to do with the ideas that the brain makes of them. "C'est l'âme qui voit, et non pas l'oeil."[23] After all, the soul is also capable of hallucinating things that one cannot see at all. Therefore, since perceptions only stimulate thinking, which then forms ideas, it is not the *res extensa,* the body with its sensory perceptions, that sees, but the *res cogitans.*

But this was already known before. The perspective view of space, praised by the Italian humanist Leon Battista Alberti, is criticized by his contemporary Nikolaus von Kues (1401–1464) in his work *De visione Dei,* which he sent to the monks of Tegernsee Monastery in 1453. He included a Dutch icon with the tract, showing Christ with the gaze of the world ruler. The monks were to conduct an experiment and stand together in front of the icon. Regardless of the distance or angle at which each individual stood in front of the image, they all felt the same gaze resting on them and found that the All-Seeing observes everything in the same way and looks at everyone and each individual at the same time. Here an absolute and unrestricted vision, a *visus absolutus,* is revealed, which is free from all restriction and far above the limited perspective view. This, according to Nikolaus von Kues, paradigmatically illustrates that man only has a much restricted vision related to a specific field of vision *(visus contractus).* One person sees more sharply, another less sharply. One person sees things close up, another sees things in the distance. In contrast, the icon suggests how one should imagine absolute vision. It also illustrates that God's absolute gaze is spatially and temporally unlimited. After all, the Greek word for God, θεός, is derived from the verb "to see" and refers to the only being that sees everything.[24]

If, as Pascal says, one sees with the soul and not with the eye, then bodies extended in space are constructions, and the *res extensa* is to be derived from the *res cogitans* and has only subordinate importance in relation to it. It has thus been shown that space, the *res extensa,* in the early modern period was subordinate to the *res cogitans.* Is this position now outdated? If one thinks of newer mathematical constructions of spaces with more than three dimensions, this does not seem to be the case. The German mathematician Bernhard Riemann presented his radically new geometry of curved multi-dimensional spaces in his habilitation lecture at the University of Göttingen in 1854, titled *Über die Hypothesen, welche der Geometrie zugrundeliegen (On the Hypotheses that Lie at the Foundations of Geometry).* Such higher-dimensional spaces are also referred to as hyperspaces. Like Francisco Sánchez, Riemann deals with extreme possibilities. He asks about space beyond the

---

[22] Descartes 1988, 704.

[23] Descartes 1988, 705.

[24] Belting 2008, 240–242.

limits of observation in the "immeasurably large" and in the "infinitely small" and introduces surfaces with constant curvature mass when he sets himself the task of "constructing the concept of a multiply extended magnitude from general concepts of magnitude. It will emerge from this that a multiply extended magnitude is capable of different measures and that space therefore only forms a special case of a threefold extended magnitude."[25] Whether in the case of hyperspace, the microcosm, the book of the world, the analogical, astronomical, or skeptical view, or in the mathematization and rhetoric in the visual arts—the activity of thinking is always emphasized, from whose designs spaces and bodies first emerge.

The Augustinian view, which wants to see the human subject more in the context of the history of salvation and eschatological events than in the context of an external physical space, can be noticed in Calderón, who probably came into contact with it in Salamanca, shaped by both Augustinianism and Thomism.[26] In his *Auto sacramental El divino Orfeo*, it is depicted how the first man, as a special creation of God, enters the world that lies at his feet.[27] Through his likeness to God, he is the crown of the world and the greatest miracle of creation. All other creatures serve him as their lord.

If, according to the book *Genesis*, the spatial world is a creation for man, serving him, then he is far superior to the spatial world. And since the interior of man encompasses the exterior and the exterior is reflected in the interior, the exploration of the spatial exterior is not necessary, as Calderón states. It thus becomes apparent that Calderón is influenced by Augustine insofar as he sees man as a subject less in his relation to external reality than primarily in that relationship to the Creator that primarily ennobles him.

The superiority of the soul over the body can also be demonstrated in the field of astrology. Astrological predictions are absolutely valid, as Calderón emphasizes in *La vida es sueño*. However, since Plato's *Timaios*, it has been valid that only the body, but not the soul, is under the influence of the stars. It is the developmental tendencies and mentality of man that can be predicted. But they only occur if the body with its passions dominates over the soul and its free will. Thus, it is also stated in *La vida es sueño*: "porque el hado más esquivo,/la inclinación más violenta,/el planeta más impío,/sólo el albedrío inclinan,/no fuerzan el albedrío."[28] The stars thus point the will in a direction, but cannot determine it as long as the soul dominates the body.

If the body is alone, it is portrayed as weak and imperfect. The same applies to the body's sensory perceptions, which appear irrelevant when they are limited to the external space. Therefore, Calderón considers not seeing, but hearing to be the most important sense. After all, as stated in Romans 10:7, faith is conveyed to man through this sense. The sight of external

[25] Riemann 1876, 1.

[26] Margraff 1912, 110.

[27] For the following references, see Margraff 1912.

[28] Calderón1978, 101.

objects, unlike hearing, does not penetrate the interior and cannot guarantee truth. Hearing allows an increase in interest and a constant enrichment of knowledge. Only hearing is capable of bridging the gap between the sensory world and the transcendent truths of salvation, between time and eternity.

In contrast, the sense of sight and touch can only grasp accidents, never substances. Substances can only be reached with the mind. In addition, the senses are subject to numerous deceptions. That the color blue is not a property of the sky, but a crystalline lie, Calderón emphasizes. Another example of deceptions is Calderón's King Baltasar in the play *La cena de Baltasar,* who is so confused by a strange appearance that he does not perceive the rest of the events around him. Optical illusions make city walls or mountains appear behind the dawn or fish to be discovered where there are none. Here, the mind is well advised to be skeptical towards external perceptions and to discern between deception and truth in sensory impressions.

Not infrequently, the senses also appear morally questionable. In Calderón's allegorical play *Los encantos de la Culpa*, Circe represents guilt and Odysseus' companions represent the senses. When they arrive on Circe's island, the sense of touch expects silken fabrics and clothes, the sense of smell sweet aromas, the sense of hearing the music of bird song, the sense of sight gold and diamonds, and the sense of taste finally the fleshpots of Egypt. After they have let themselves be beguiled by Circe's ladies, the sense of sight by envy, touch by lust, smell by slander, taste by gluttony, and hearing by flattery, they are transformed into animals: the sense of sight into a tiger, the sense of touch into a bear, the sense of taste into a pig, the sense of smell into a lion, the sense of hearing into a chameleon. In Calderón's *Autos sacramentales*, all senses thus seem susceptible to the temptations of immorality. The allegory of lust in *El año santo en Madrid* aims to appeal to the eyes with her beauty, to the sense of hearing with her voice, to the sense of touch with her softness, to the sense of smell with her fragrances, and to the sense of taste. So if one has to be wary of what the senses offer as perceived, should the senses then be completely rejected as something negative?

The allegorical figure of man in *Los encantos de la culpa* argues against this and suggests that the senses can at least be credited with the function of distracting from the hardships of everyday life. However, the figure of reason does not want to accept this and argues that without hardships in this world, there would be no joy in the next. Here, therefore, this world is contrasted with the next, and the equation is made that joys in the next world are made possible by hardships in this world. If the senses seduce and distract man, i.e., make him guilty, then he should free himself from them and the guilt associated with them as much as possible. The figure of reason advises countering guilt with penance.

In the early modern period, however, seeing also gained a new appreciation. Juan Luis Vives had in his writings *De causis corruptarum artium* (1531) and *De tradendis disciplinis* (1531) argued against the derivation of knowledge from the writings of antiquity and for observation and experience, before John Locke (1632–1704) and the Anglo-Saxon empiricists and sensualists saw observation and experiment as sources of knowledge. Their investigative gaze is no longer contemplative, but focuses on the obvious and evident, which the eye can finally see, supported by the new inventions of the microscope and telescope.

Indeed, Calderón does not lack appreciations of seeing either. After all, in *La nave del mercader*, creation is presented as the spectacle of a divine author with the motto "Live to see," with the actors being the four elements. The beauty of perspectives and appearances, which are brought about by changing weather conditions over time, is evoked. But here too, the restriction in case of misuse is not missing. Despite all the beauties, the great mass has nothing better to do than to seek love affairs and jealousy. In everyday life, seeing is even superior to hearing, as the eyewitness is more trustworthy than the reporter. Here, the competition and rivalry between the sense of sight and the sense of hearing become clear, which have a long tradition behind them.

In the plea for hearing, one can refer back to the Gospel of John: In the beginning was the Word, and the Word was with God, and God was the Word.[29] Luther explicitly prefers hearing when he says that the kingdom of Christ is: "a hearing kingdom, not a seeing kingdom. For the eyes do not lead and guide us to where we find and get to know Christ, but the ears must do that."[30] Calderón does not refer to Luther, but to the Old Testament, when he has Moses say in *El viático cordero* that faith reaches the sense of hearing through the word and that no faith is possible where only seeing is present. Hearing appears superior to seeing also because it does not stick to the outer appearance like seeing, but encounters the inner reality. This is formulated in an Aristotelian way in *La divina Filotea*, where the allegory of reason assigns to seeing and feeling the externally changing properties, but not the essence of things. In *El valle de la zarzuela*, hearing seems to be more of an inner sense than seeing. This becomes clear from the hint that one can see a beauty over and over again without it changing, the image thus remaining the same and no gain in knowledge being possible. It is different with hearing. Repeated hearing provides confirmation and thus progress in knowledge.

That hearing gains its dignity through its connection with faith becomes clear in *El nuevo palacio del retiro*, where the senses argue over a bouquet held by the allegory of faith. They highlight their advantages, as in the already mentioned *Los encantos de la culpa,* in distraction and in the mediation of worldly pleasures. Only the figure of the sense of hearing emphasizes its imperfection. But faith turns to it and hands it the bouquet, as it is characterized by the ability to perceive faith. That faith is associated with the sense of hearing is also evident in *La Iglesia sitiada*, where the figure of hearing

---

[29] Gen 1:1; Joh 17:5; 1 Joh 1:1–2; Ap 19:13.
[30] Luther 1883–2009, 11.

appears as a silent companion to the figure of blind faith, who responds to heresy that she can hear what is sufficient for faith, while heresy wants to see in order to believe.[31]

If hearing is attributed with not clinging to external appearances, that unlike seeing it is not oriented towards worldly pleasures and accidents, but towards substance and the increase of knowledge, then in Calderón hearing seems to be more of an inward activity and less to do with perceiving sounds and noises. When Calderón then connects hearing with Christian faith, the acoustic level is merely an allegory illustrating the reception of meaningful content. Such allegorization also takes place where the inner senses are spoken of, which are tasked with compensating for the weakness of the outer senses.

The systematization of the doctrine of the inner senses goes back to Origen and Bonaventura.[32] The Spanish humanist Alejo de Venegas (1497–1562) calls them "ojos del alma", eyes of the soul.[33] An example of the outer senses being used as metaphors for the inner senses is taste. Nourishment and recognition are paralleled when we speak of thirst for knowledge or hunger for recognition. The Latin word for wisdom, "sapientia", is derived from the verb "sapio", I taste.[34] This verb in turn is used in Latin in a figurative sense and means "I judge correctly," which transfers from tasting to taste, understood for example as judgment formation in aesthetics. For the Christian philosopher Ambrose, the outer eye serves for seeing *(videre)*, the inner eye for recognizing *(recognoscere)*. Unlike animals, humans can see with their inner eyes even with their outer eyes closed, and the former are not dependent on space or time, nor on sources of light.[35] Augustine distinguished between the outer and the inner man, with the latter being understood in analogy to the former. While the outer eye is susceptible to sins in a Christian view, the inner eye can fathom the secrets of the soul or the divine. According to Platonic tradition, truth is recognized only with the inner eyes, the eyes of the spirit. The inner eye is thus nothing more than a metaphor for *intelligentia, memoria, intellectus,* or *ratio.* Synonymous with *imaginatio,* it allows us to imagine a spring day in winter. The inner eye also enables the mystical vision of God. Since antiquity, it has been referred to as *oculus mentis,* which in Augustine is the same as *oculus cordis.* In Augustine, God himself becomes the preferred object of vision for the purified eye of the heart.

Here, the proximity of the inner senses, especially the inner eye, to mystical vision is indicated. Indeed, the mystic Teresa of Ávila (1515–1582)

---

[31] Cf. Margraff 1912.

[32] Schrader 1969, 114.

[33] Schrader 1969, 114.

[34] Von der Lühe 2007, 340–355.

[35] Schleusener-Eichholz 2007, 368–375.

refers in her visions to the *visio spiritualis* or *imaginaria* and speaks of the eyes of the soul and the inner senses.[36] Juan de la Cruz (1542–1591) relates the inner senses to the mystical experience and sees their effect in visions and revelations of the mind, in which physical perception is not involved in any way.[37] For Ignatius of Loyola, the founder of the Jesuit Order (1491–1556), the practitioner should direct his inner eyes, the "sentidos de la imaginación", first to hell, then to heavenly things, so that autosuggestive images replace the divine inspiration of mysticism.[38] Ignatius' predecessor Thomas of Kempen (1380–1471) succinctly praises the eyes that are closed to the outside but open to the inside: "Beati oculi, qui exterioribus clausi sunt interiorius autem intenti."[39]

That Calderón gives priority to the mind over the senses is also evident in *Andrómeda y Perseo*, where he compares human nature with the beauty of the heavens and lets man assert that he is ultimately a rational being with senses and soul.[40] But what does this rationality consist of? The "inner senses" can refer to such different abilities as *intelligentia, memoria, intellectus, ratio,* or *imaginatio.* Now all intellectual activities do not belong to the body, but to the soul. Therefore, Calderón can replace the figure of the *Entendimiento,* or understanding, with the soul. If the senses are oriented towards matter, the soul owes its form to the heavens.[41] What the senses, as part of the material body, have experienced of material things, the soul should process.[42]

In summary, it can be stated that the negative ethical evaluation of the senses in the *autos sacramentales* is reinterpreted in a Christian way by devaluing the external and worldly, and upgrading the internal and otherworldly. The fact that extended reality *(res extensa)* is less important than the mind *(res cogitans)* was introduced by Plato through his Theory of Ideas, according to which the perceptible world depends on transcendental ideas. In his allegory of the cave, he described the path of the wise from apparent phenomena to ideas and to the highest idea, which becomes a Christian pilgrimagee for the Spanish author Fray Luis de León. In Augustine, in the early Middle Ages, the highest idea is equated with God and a *civitas Dei* is distinguished from a *terrena civitas.* Because God, as the world's architect, has applied the same arts that man uses in the exploration of nature, geometry and mathematics are helpful. However, sensory perception can only grasp accidents; the substance is reserved for inner perception by the mind or soul. The relativity

---

[36] Schrader 1969, 116.

[37] Scheerer 1995, 841.

[38] Schrader 1969, 121.

[39] Schrader 1969, 113.

[40] Calderón 1995, 151, 19–20.

[41] "Que si de la tierra han sido/los sentidos, porque ella/de su materia los hizo,/el cielo ha de dar la forma/al alma." Calderón 1996, 110, 718–723.

[42] "Es preciso/que si a la tierra te obligas/a volver lo recibido/de la tierra, que es el cuerpo,/ hayas de volver lo mismo/al cielo cuya es el alma." Calderón 1996, 110, 725–730.

of seeing is illustrated by Nicholas of Cusa with his concept of unlimited seeing and Bernhard Riemann with his hyperspace. Calderón assigns sensory perceptions to the body and sees them exposed to the seduction arts of immorality. Distinguishing between this world and the next, he assigns a higher value to hearing than to seeing, as faith is conveyed through the heard word, while seeing remains attached to the external appearance. The *ojos del alma*, the inner senses, however, open up mystical visions and Jesuit contemplations. The preference for the inner world and the limitation of observing the outer world would be completely unsuitable on the side of the antipode, where the sensualistic and empirical approach of a John Locke stands, in which knowledge acquisition starts from the experience of the outer world.

3.3   Teleology

"Teleology" refers to the view that all actions, events, and developments are purposeful and aim at a purpose. The Greek word *télos* means "fulfillment" or "achievement", so it has something normative. The corresponding Latin *finis* is initially thought of as a boundary marker. The underlying thought is that everything has a well-defined place, its boundaries, but also its purpose of existence. In Plato's *Phaedon*, Socrates asks the question of the purpose of existence, the good or best, through which things receive their meaning. Actions are not causal processes, but are determined by a what for, by a good, therefore, that the actor strives for. In Plato's book about the state, the *areté*, or competence, determines all being, beyond the sensual and recognizable things. To the purposeful world, Plato opposes a demiurgic creator who wants everything to be good. The word *skópos*, used in archery, refers to hitting the right target and is applied by Plato to the goal of life. Aristotle relates the *télos*, the goal, to action and nature. Thus, the swallow does not build a nest by chance, but for the purpose of its self-preservation. Plants do not drive their roots into the air, but into the earth with the goal of protecting themselves. An inner tendency towards one's own perfection causes the goal-oriented activity, through which the respective definiteness and essential form is achieved. In ethics, happiness is the right goal, which is achieved through virtue.

The theory of forms does not allow for the development of species from one or more original species. Just as the arrow strives for the target without knowing it, so according to Thomas Aquinas the unconscious nature also strives for its goal, guided and directed by the creator. The activity of animals always remains the same: the swallow builds its nest, the spider weaves a web. Purposefulness in the animal kingdom exists in instinct. The animal owes the preservation of its life and that of its species to the drive for food and sex. All beings carry an immanent purpose in themselves: They should perfect themselves and bring the possibilities inherent in them to full development. If one asks about the consequences of the final cause for society, it is best to look directly at its foundations. According to Aristotle, the goal of man is *eudaimonia* (felicity), which can only be achieved in the city-state. Therefore, he concludes, man is, from his goal, i.e., from his nature, a social being.[43]

Perfection means the agreement of is and ought, which can be meant in an ontological, ethical, and aesthetic sense. In Plato, the idea of the good is the measure of all being. The sensual being never has the perfection of the idea, but always a striving for perfection. In Aristotle, the perfect is first defined as that which does not have a single part outside of it; secondly as that which cannot be surpassed in terms of its ability (*areté*) (e.g., a doctor or a sophist); thirdly as that which has achieved its good goal or purpose and is thus completed.[44] Building on this, in the High Middle Ages, Thomas Aquinas sees the *perfectio prima* given when a thing is perfect in its substance under completeness of parts. A second perfection refers to the goal, a third also includes the accidents.

Since Thomas Aquinas shaped medieval scholasticism, he is presented in detail. What in Plato was the highest idea, is in him the highest purpose. He sees nature and man oriented towards the highest purpose. "The purpose is the general category under which the entire universe as well as the life of the individual is considered."[45] Because Thomas follows Aristotle, he also sees every action directed towards that goal, which Aristotle defines as "the 'what for'". One takes a walk to become healthy. The house that the architect wants to build is the goal of his will activity, to which he subordinates other things. Everything that happens because of a goal is influenced by it. The goal, as *causa finalis*, has a causality. To realize a possibility, an act, an active principle is required. As with Aristotle, Thomas does not always distinguish precisely between principle and cause, i.e., *arché* and *aitia*. The principle, *arché*, is defined as the first in an order, without this actually having to exert an influence. The final cause, *aitia*, on the other hand, pre-exists as a thought in the consciousness of man and is at the beginning of the consideration whether something should be done; only then do the means come into play.

The will can only be influenced by what appears valuable or pleasant, thus as good. The good is defined in Aristotle and Thomas as that which everything strives for.[46] That for which something happens should be the best and the purpose of the rest. Therefore, evil cannot be the goal. However, something that is objectively evil can appear subjectively good.

Thomas follows Aristotle when he brings the Platonic idea from the beyond into reality, where he sees it determining things through formal cause and final cause. It thus forms two causes in the Aristotelian four-cause doctrine, where he sees the formal and the final cause as coinciding. The soul

---

[43] Aristotle 1995d, 4.

[44] Aristotle 1995a, 113–114.

[45] Steinbüchel 1912, 1.

[46] Steinbüchel 1912, 27.

in the organic body is for him the formal cause, moving principle and purpose of the body. The final cause stands first in the system of causes as *causa causarum*. Since God and nature do nothing without a purpose, a definition must not merely state what is given; it must also state the goal causes that determine it.[47]

The idea of the goal has consequences for Thomas' conception of society and the cosmos. Man as a social being (*zoon politikon*) is dependent on others. The basic condition is the subordination of the many under one ruler, who has the common good in the state as a goal.[48] Further goals of the state are the maintenance of order internally and externally, material well-being of its citizens as well as the cultivation of intellectual and ethical goods. When Thomas conceives the universe teleologically, man is the culmination point, as the less developed always serves a higher goal, thus man. The parts of the universe are there because of the whole. The world appears as an ordered army, over which the commander, i.e., God, stands as the ultimate goal and highest good. Therefore, creatures strive for similarity with him through imitation. The ultimate goal of all things is to become similar to God.

The English mathematician and political theorist Thomas Hobbes (1588–1679) criticizes the Aristotelian conception of man, whose goal is community.[49] Instead, ambition, competition, and scarcity of goods ensure that everyone is a wolf to the other.[50] In Hobbes, the natural state thus proves to be morbid and something to be overcome. Hobbes also has a different conception of felicity than Aristotle. For him, it does not consist in the peace of a contented mind: For there is no such *finis ultimus* (ultimate goal) or *summum bonum* (highest good) as is mentioned in the books of the old moral philosophers [Aristotle is meant]. Happiness is a constant progression of desire from one object to another, the attainment of one being always only the way to the next.[51] Hobbes therefore does not start from the goal of a good life, but from bare survival. People are comparable to mushrooms that have sprung from the earth, without one being obligated to the other.[52] Here we see a conception completely contrary to Aristotle, where the goal, felicity, is not inherent in man as a goal, but safety can only be achieved through contract or state power.

However, there are just as many proponents as opponents of teleology. The physicist Isaac Newton(1643–1727) argues that the natural scientist can

---

[47] Steinbüchel 1912, 37.
[48] Steinbüchel 1912, 104.
[49] Wolfers 1991, 61.
[50] Hobbes 1996, 104.
[51] Hobbes 1996, 80; Fetscher 1960, 683.
[52] Hobbes 1966, 82–83.

recognize the wisdom of the creator precisely in the final causes. Gottfried Wilhelm Leibniz (1646–1716) adds that in physics, mechanical causal causes always also depend on final causes; since there is nothing but activities, every activity is self-activity, and thus all reality is to be seen as purposeful activity of the involved things, which he calls "monads".

In contrast to natural science, which has nature as its object, literature deals more with humans, which is why its criterion is relevance to the subject and not progress. The freedom of the purposeful acting subject is opposed to the determinism of causality in nature. Teleological thinking also dominates the medicine of the *Siglo de Oro*. Since in Spain there was not yet the strict separation of the body as *res extensa* from the soul as *res cogitans* that Descartes had made in 17th century France, the relationship between body and soul plays a major role. The Neoplatonic conception of nature as a product of the mind and of the correspondences between microcosm and macrocosm dominates, resulting in numerous dependencies and harmonies between body and soul or body and mind. Plato had already spoken of the world soul as the force that moves itself and everything else, and had contrasted it with the individual soul, while for Aristotle the soul of plants is the power of nutrition. In the souls of animals, the sensations of pleasure and displeasure are added, while man possesses reason in addition to the powers of plant and animal, which is of divine origin. Since the human soul thus unites the powers of all other beings and of the divine within itself, it can be seen as a microcosm according to Aristotle. With humanism, the Neoplatonic conception spreads, according to which nature is a product of the mind, so on the one hand the deity unfolds in nature and on the other hand the deity is the highest unity point of the different sciences. Against this background, natural science becomes theosophical. When Plotinus speaks of the beauty of the macrocosm, this beauty is also seen as the appearance of the divine idea. If everything has a cause and the last formal and effective final cause is God, then the universe is the essence of God made creature and is to be considered pantheistically. God becomes the unity in which all opposites are abolished, the *coincidentia oppositorum*, and acts as *natura naturans*, shaping and explicating the *natura naturata,* the universe and all creatures. The idea of creation leads to the notion of intention and purpose. Where purposes underlie, teleological thinking is applied.

Francisco Diaz, the physician and surgeon of King Philip II, therefore lets in his *Compendio de cirurgia y anatomia, en el qual se trata de todas las cosas tocantes a la theorica y pratica della, y de la anatomia del cuerpo humano, con otro breve tratado de las quatro enfermedades* (1575) an intern ask about the purpose of the gallbladder located next to the liver. In his explanations, it becomes clear that any malfunction of an organ affects the functioning of the other organs. The bones, as the skeleton, form the foundation, which has the purpose of supporting the other body parts. They are like the element of earth, cold and dry. Apart from the teeth, they have no sensations.[53] The special purpose of the eyes is highlighted, "la excelencia de su oficio para lo fueron constituydos."[54] Therefore, they are round so that one can look everywhere, and the pupil reflects things so that the eyes can see things as they are. In addition to the other components, the seven muscles that help the eye move in all directions are also described.

The magnet, which has the power to move from within itself without being moved, is for Thales in antiquity a model of the soul. In Plato, the body becomes a test, a prison for the immortal soul. For Aristotle, the soul is the purpose, form, and cause of motion for the body, from which it cannot be separated. Aristotle defines the soul as follows:

> "The soul is the cause and principle of the living body. However, this is understood in several ways. According to the three distinguished types of causes (principles), the soul is likewise (triple) cause: It is both the origin of movement and purpose, and also as the essence of animate bodies, the soul is cause. [...] All natural bodies are organs of the soul, and like the (bodies) of living beings, so are those of plants for the sake of the soul. The purpose is of twofold meaning, one as for-what, the other as for-which."[55]

So if the soul acts as cause, purpose, and essence (form) on the body, then this can lead to health or disease for him. Body and soul form a system in which the soul acts as final cause and the body as effect. The holistic view with a dominance of the soul over the body is shared by Juan Luis Vives (1492–1540). He too speaks of the soul as form and purpose of the body[56] and sees it as an active principle, since, when it undertakes something with the tools of the body, the activity originates from it.[57]

Following Aristotle, the representative of the School of Salamanca Francisco Suárez (1548–1617) defines the soul firstly as rational and independent of matter and equipped with intelligence and will, and secondly as the formal cause of the body with its material activities.[58] For the body, it is the active element. Also in Suárez, the rational soul becomes the formal cause for the body and its movements, where form is nothing other than goal and purpose.

In 17th century Spain, the strict separation of outside and inside is not made, as a look at the presentation of the different medical approaches of Juan de Cabriada in his 1687 published *Carta filosófica, médico-chymica* shows. In classifying diseases by types and causes, he refers to the different schools. For Paracelsus, as he reports, there are diseases that God assigns as

---

[53] Diaz 1575, 58v.

[54] Diaz 1575, 67r.

[55] Aristotle 1995c, 37.

[56] Vives 1945, 53.

[57] "Así, en el pintor está la facultad de pintar, y en mí la de escribir." Vives 1945, 49.

[58] Suárez 1978, 19.

punishment for sins, those that are conditioned by the course of the stars, others that have a flaw in nature as causes. In addition, there are those that are caused by imaginations and passions of the soul, and others that arise from ingesting a poison.[59]

In the *Nueva filosofía de la naturaleza del hombre* (1587) by Oliva Sabuco de Nantes, emotional happiness, based on knowledge of the world and the choice of the right measure, is a prerequisite for physical health. Most of the advice Sabuco gives for maintaining health concerns the soul, which should keep itself free from negative affects. Anger, for example, is compared to a wicked and dangerous animal that one must confront in order not to suffer a much greater damage than the one that causes the anger, namely the loss of health.[60] The same applies to fear or worry about the future, which can be damaging to health or even fatal. Fear is often more dangerous than the event one fears, as it causes harmful melancholy with its false illusions and suspicions.

Just as negative affects of the soul damage the health of the body, positive affects promote it. The more negative the affects of hopelessness, the more positive the affects of hope.[61] Hope and joy invigorate, harmony and friendship let soul and body thrive. In Sabuco's opinion, happiness, pleasure, satisfaction, and joy are the purpose of human life, which are also responsible for health.[62] In her health-promoting behavioral rules, Sabuco advises the soul to maintain the virtue of moderation and thus the golden mean. Sabuco recommends moderation, especially in food intake. From the ancient doctors, the sentence is handed down that more people are victims of gluttony than of the sword.[63]

Hieronymo Merola's 1587 published *Republica original sacada del cuerpo humano* relates the structure of the state as well as human activities and sciences to man as a microcosm. The relationships that Merola, as a doctor of philosophy and medicine, establishes are diverse. For example, the stars correspond to the five senses and the four elements to the four temperaments.[64] The contemplative sciences are there for the soul and the active sciences for the body. Those who limit themselves to contemplation are unsuitable for government affairs. Everyone should align themselves with the common good, some as commanders, others as obedient. The same applies to the microcosm of man, who is the final cause of the macrocosm and should therefore find his way back to the creator through virtue.[65] Virtue is the

---

[59] Cabriada 1686, 105–106.

[60] Sabuco 1981, 90, 88.

[61] Sabuco 1981, 129.

[62] Sabuco 1981, 123.

[63] "Más mata la gula, que la espada." Sabuco 1981, 115.

[64] Merola 1587, 118a.

[65] "Que el mundo pequeño que es el hombre, es final causa a la qual se refiere el grande: y Dios es fin del grande y del pequeño. Porque lo que pretende el hombre es hazer una circulación y volverse a Dios de quien tiene su origen, y esto mediante la virtud, con la qual viene a hazerse tan virtuoso, tan perfecto, y semejante a Dios, que por la similitud es atrahido por el summo bien." Merola 1587, 20b, 21.

origin and goal in the microcosm as well as in the macrocosm for Merola. Here again, a Neoplatonic, pantheistic background is evident. Not following the virtues would miss the purpose.

However, Epicureanism and Stoicism offer different goals for life fulfillment. While the followers of Epicurus(34–270) advocate the intensification of pleasure, the school of Stoicism proposes the suppression of passions, as they hinder reason. In France, Epicurus is celebrated in the 17th century by Pierre Gassendi as a master of the art of living, adding a materialistic component to his anti-scholastic and scientific view. It was the Italian Lorenzo Valla who, with his treatise *De vero falsoque bono* (1433/4), originally titled *De voluptate* (1431), presents a confrontation of Epicureanism with Stoicism, seeing Epicurus under Christian auspices as the worldly mode of happiness as a precursor to the otherworldly form of being.[66] For Valla, pleasure is the goal, which consists in the joy of the mind and body. The interest in self-preservation and pleasure refers to the self, making others secondary. The primary is one's own advantage, then comes that of relatives and finally that of the homeland. Here, the ethics oriented towards the common good, the *bonum comune*, is reversed. False is the sacrifice of self-surrender, which leads to death in war for the homeland. It is also wrong to sacrifice oneself for others. Behind many deeds praised as heroic is nothing but the desire for fame. But even if Valla's own well-being dominates, the well-being of others remains connected to it as the enjoyment of community. However, when Valla puts his Epicureanism into the perspective of Christian ethics as a soteriological ethic at the end of his explanations, he sacrifices the original earthly right of pleasure to a Stoic virtue theory in the service of Christian religiosity.[67]

The ethical teachings of the Stoics assume that a successful life is only possible in rational harmony with nature, which is determined by a rational principle of world events. From this, the goal of life according to nature, the *secundum naturam vivere,* is derived. Happiness is only achievable through virtue and virtuous action. Vices are something bad; honor, power, beauty, and health, on the other hand, are indifferent compared to the highest good of wisdom. Affects such as pleasure, pain, fear, or desire are to be avoided in favor of an unshakeable and confident attitude. Reason, as the bond between people, creates community and leads to the idea of a pantheistic, reason-based world of gods and humans. In the early modern period, Neostoicism gains great influence. The Dutchman Justus Lipsius drew on Seneca and Epictetus in his works *De constantia* (1584), *Manuductio ad Stoicam philosophiam* (1604), and *Physiologia Stoicorum* (1604), where the Stoic doctrine appears to him as a prefiguration of the truths of Christianity. *De constantia* was written under the impression of the Dutch revolt against Spain and aims to

[66] Leinkauf 2017, 709.
[67] Leinkauf 2017, 719–721.

offer citizens support and comfort in the face of political-religious conflicts. Lipsius' orientation towards the Roman Stoa seems more suitable to him to bring peace and unity than the numerous contemporary controversial theological writings. Lipsius was convinced that those acting politically could overcome the crises of the time with his teachings and even prevent them in the future.[68] In *De militia romana* (1596) he recommends taking the Romans as a model in military matters, whose discipline he believes is lacking in the soldiers of his time.[69]

It thus becomes apparent that where man is placed at the center, human action which is goal- and purpose-oriented becomes paradigmatic. It becomes a model when it comes to describing the process of knowledge. Things gain their meaning not through causal connections, but through a what for. Purposes of existence determine the unconscious nature as well as the ethical action of man, who aims for the goal of felicity, which he as a social being can only achieve in a state whose goal is the common good.

Parts of the universe exist because of the whole, which is their purpose. Purposes also operate in individual things. Thus, bones have the purpose of supporting the rest of the body parts. Eyes are round because they have the purpose of looking everywhere. Just as the architect's plan sets the goal and becomes the final cause of the house, so goals are the actual causes everywhere. Just as the microcosm of man is the final cause of the macrocosm, so the soul is the purpose of the body. If the soul and mind miss their purpose, if they are therefore unhappy, then the body becomes sick. However, Epicureanism and Stoicism offer different paths to the goal of happiness. Regardless of whether human action itself is considered or its model is transferred to the observation of nature, it is the final causes that are given greater importance than the causal causes. This is also related to the preference for the inner world over the outer world. As already mentioned, the Platonic idea of Aristotle and Thomas Aquinas became the form and final cause, thereby determining the substance of a thing. This, however, is more important than its accidents accessible to external observation.

3.4   EXCURSUS: ROMANTIC PHYSICS

The philosophy of German Idealism, as represented by Fichte, Schelling, and Hegel, gave priority to the subject over the object. The opinion was that in knowledge as in action, the subject is the starting point. Just as the absolute spirit has created the world, so the individual subject creates its reality from within itself and with its mental creative power. Activity and design therefore

---

[68] Mout 2012, 118.

[69] Mout 2012, 136.

originate from the subject, so that the realization of the object is due to the subject. Schelling, who like Johann Gottlieb Fichte, August Wilhelm Schlegel, and Friedrich Schlegel, was involved in the early Romanticism of Jena, formulates it this way: "God is that which presupposes no other concept, like space in geometry. The world, however, can only be understood as a consequence of him."[70] The divine will becomes the beginning of nature, thus creating nature. So God is that cause in the world process, "which gives the ideal precedence over the real."[71] What applies to God is also true for creatures by analogy: "The finite beings, on the other hand, only have the freedom to eternally set themselves."[72] This means that reality explicitly represents what is implicitly already given in the idea. Nature arises by the moments contained in the idea falling apart and existing side by side. It is nothing but the idea that has fallen apart, "for every being only realizes itself by setting the implicit being as explicit."[73] The inner world therefore has priority over the outer world for Schelling:

> The true fact is always something internal; the true fact of a battle, for example, lies in the mind of the general, not in the attacks or cannon shots; the true fact of a book is known only to the one who understands it.[74]

At the time of Romanticism, philology and physics were parts of the Faculty of Philosophy at German universities. Therefore, a comparison of the development of the two subjects will be undertaken in the following. First, some characteristics of national philology at the time of Romanticism will be highlighted. Then, the path of physics up to Romanticism and its conception in Romanticism will be briefly outlined. In the 19th century, the new philological subjects at the universities, like the natural sciences, emerged from the Faculty of Philosophy. While natural science faculties separated and gained independence in the first half of the century in other European countries, this did not happen in the German-speaking area until the second half. When Leiden established a natural science faculty in 1811, Amsterdam in 1815, and Louvain in 1834, Vienna followed in 1872 and Frankfurt in 1912. Prior to this, Zurich had divided the Faculty of Philosophy into a philosophical and a natural science section in 1859, Munich in 1865, and Würzburg in 1873. Physics now divided into mechanics, thermodynamics, molecular physics, electricity, acoustics, and optics.[75]

---

[70] Schelling 1989, 68; cf. Strosetzki 2022, 271–279.
[71] Schelling 1989, 97.
[72] Schelling 1989, 106.
[73] Schelling 1989, 53.
[74] Schelling 1989, 37–38.
[75] Weber 2002, 211–213.

Why are German-speaking countries latecomers? This is not least due to the influence of Romanticism and the philosophy of German Idealism on the Faculty of Philosophy. The Romantics, like the representatives of the philosophy of German Idealism, equated the productive force of the absolute spirit with that of the individual spirit and the spirit of the people. They thus preferred ideality over reality. Precisely for this reason, Don Quixote, with his striving for infinity, is a suitable modern myth for them. Just as according to Friedrich Wilhelm Joseph Schelling the absolute spirit has created the world, so the individual subject creates its reality from within itself and with its intellectual creative power.

In scholasticism, theology and philosophy were connected into an entire system,[76] so that physics, like philology, had its place in the basic philosophical studies and was linked with logic and metaphysics. In Aristotle, every thing strives for the perfection of its essence, its form. Therefore, natural processes, like ethical actions, are equally final. Although there were alchemical laboratories in the 17th century and mechanistic Cartesianism emerged, physics at the universities was Aristotelian and was taught in late scholastic tradition, with the school philosophers seeing themselves not as researchers, but as transmitters of a tradition that was only about order, justification, and preparation of materials. Physics, like philology, works with traditional texts, is contemplative, and not aimed at controlling nature. At the Catholic universities, the Aristotelian tradition continued until the mid-18th century, while the mechanistic tradition established itself earlier at the Protestant universities through the mediation of the Enlightenment thinker Christian Wolff. Because the mechanistic method was viewed critically for religious reasons due to its lack of teleology, efforts were popular to combine, in the sense of Leibniz, the mechanistic method with the Aristotelian one, i.e., a mechanistic physics with a teleological view of nature.[77] By the end of the 18th century, the paradigm shift from Aristotelian to mechanistic physics was complete, and thus the cosmos, which had been meaningfully ordered according to a divine plan, became a world machine with purposelessly caused events. The later change from mechanical to inductive and experimental physics of the forces in Newton was no greater, as both are concerned with the changes of bodies as movements. Natural philosophy became natural science and transformed the Aristotelian question: "Why is nature as it is?" into the question: "How does it work?" This corresponded to the statement by Karl Marx: The philosophers have only interpreted the world in various ways; the point, however, is to change it. For now, through mathematical descriptions and constructions, the steam engine, the railway, and the radio were created.[78]

---

[76] The second scholasticism encompasses philosophy and theology, with the former having a propaedeutic character for the latter. Ramis 2024, 36.

[77] Lind 1992, 54–58.

[78] Coy 2003, 1.

As already explained above for philology, Romanticism and the philosophy of German Idealism also brought a new impulse for physics. By again starting from philosophical and literary premises, they take a step back for a short time. Based on the idea of the unity of man and nature, the task of understanding nature is to bring man closer to nature, as a look at the *Grundriß der Experimental-Physik* by Karl Wilhelm Gottlob Kastner(1783–1857), professor of physics and chemistry at several German universities, shows. The work, published in 1810, was intended for use in basic lectures at the university.[79] Kastner, along with Friedrich Schelling, assumes that nature is a manifestation of the divine spirit, that nature and spirit are one and the same, and that the human spirit participates in the spirit of nature. Every physicist strives to rediscover his own laws of life in the life of the whole. He approaches nature intellectually to explore its spirituality. In doing so, he does not make nature a mere object, but seeks the idea of the whole, which is the true, beautiful, and good in nature. In this search, the researcher stands alongside the artist and the priest. Nature is not only an object of knowledge, but also of admiration and worship. Romanticism opposes the metaphor of the machine with the organism or the world soul. All organisms have a soul, so nature is a dialectical unity of productivity and product, in the sense of Schelling's *natura naturans* and *natura naturata*. Kastner even applies this idea to inorganic nature and speaks of "inorganic life" or "anorganisms". The history of nature is defined teleologically as the determination of the higher development potentials of the spirit. Evolution theory becomes ideal genesis and nature is nothing other than the representation of the activities of the primal spirit or God. Thus scientific education in nature teaches us to see the whole and absolute.

Even in the 18th century, the educational goal of the secondary school was the savant, who had acquired a certain knowledge and demonstrated in disputes that he could deal with scientific literature.[80] At the beginning of the 19th century, the Faculty of Philosophy, as already mentioned, had emerged from its propaedeutic tradition of the *artes liberales* and had become an equal faculty, in which physics was one among other subjects. The professor of physics initially held only a one- to two-semester basic lecture in addition to private seminars. Even in 1860, in the eighth edition of his widely disseminated work *Lehrbuch der Physik zum Gebrauche bei Vorlesungen und zum Selbstunterricht*, the physicist Wilhelm Eisenlohr (1799–1872) does not entirely abandon the teleological method when he wants to be understood under "nature": "Partly the sum of all sensually perceptible things, the whole physical world; partly the totality of all properties, forces and relationships of a thing; but also the first cause of all things."[81] After highlighting the

---

[79] Lind 1992, 279–295.
[80] Lind 1992, 314–324.
[81] Eisenlohr 1860, 1.

influences of physics on industry and the wealth of nations, he addresses the positive effect of physics on religious and moral feeling: "Through it we learn to admire the wisdom and greatness of the Creator everywhere, as we learn how the most diverse and wonderful purposes are achieved through the application of the simplest means, and what spirit of order, harmony and power pervades the entire universe."[82] He then deals with bodies, wave motion, sound, light, heat, magnetism, electricity, and electrodynamics.

Georg Simon Ohm in his book *Grundzüge der Physik als Compendium zu seinen Vorlesungen* intends to arrange his material taking into account the sparse mathematical knowledge of his students, "who mostly come from the humanistic educational institutions."[83] In the division of natural science, he distinguishes between inner world and outer world and inner thing and outer thing or inner and outer nature, the latter being the subject of natural science and the former the subject of soul science. He also does not fail to start from a common cause: "Both the inner and the outer nature flow together into a common origin, in the world creator incomprehensible and unfathomable to us worms."[84] As romantic natural philosophy gives way to positivism, however, teleological thinking and with it the commonality of physics and philology, newly formulated in romanticism, disappear.

Romantic physics thus once again draws on the early modern models, which it finds in the philosophy of German Idealism in a new formulation. It sees a parallelism between the absolute spirit that created the world and the individual subject that creates its reality from within itself. Nature becomes a manifestation of the divine spirit and its history is characterized teleologically as the higher development of the spirit. Romantic physics is thus a special development based on the philosophy of German Idealism, which can also be observed in medicine in 19th century Germany.[85]

[82] Eisenlohr 1860, 3.
[83] Ohm 1854, III.
[84] Ohm 1854, 1.
[85] Engelhardt 2023.

PART II

UNCLEAR PRIORITIES

CHAPTER 4

Fields of Tension

Literature and natural science can be characterized and evaluated by their different features. While intuition, rhetoric, and tradition are on the side of literature, natural science is often characterized by systematics, logic, and progress. When these characteristics are demonstrated by different authors in the following, it will become apparent that a clear idea of the primacy of one side or the other has not yet emerged.

4.1   INTUITION AND SYSTEMATICS

Even in the question of how the available knowledge should be presented, literature and natural science diverge. It is not just a question of arrangement. Because depending on this, the style is also different, as is the approach. The mathematician, physicist, and writer Blaise Pascal proposed in the 17th century in France to distinguish between an *esprit de géométrie* and an *esprit de finesse*. While the former prefers general principles but moves outside of everyday reality, the latter prefers the everyday world. The geometer lacks finesse because he does not see what is before his eyes and is accustomed to dealing with the clear but coarse principles of geometry and can only arrive at results from them.[1] In contrast, the *esprit de finesse* does not make principles, but everyday observation the starting point. One should be able to see something spontaneously and directly at a glance, not through progressive thinking.[2] Discursive thinking is thus opposed to intuition. The two seem to be incompatible. Because geometers make themselves ridiculous with their definitions and derivations in the field of *finesse*, and the *esprits fins* lack the

---

[1] Pascal 1963, 576.

[2] "Il faut tout d'un coup voir la chose, d'un seul regard et non pas par progrès de raisonnement, au moins jusqu'à un certain degré." Pascal 1963, 576.

© The Author(s), under exclusive license to Springer-Verlag GmbH, DE, part of Springer Nature 2025
C. Strosetzki, *Literature in Dialogue with the Natural Sciences*,
https://doi.org/10.1007/978-3-662-71319-8_4

patience to reach the first principles through speculation, which are therefore only accessible to the *esprit de géométrie*.

A universal language oriented towards the geometric-mathematical model is envisaged by Descartes in a letter to Mersenne in 1629, but he doubts its feasibility.[3] The problem of separating discursive "geometric" thinking and intuitive perception of reality intensifies with Descartes, who distinguishes between a *res cogitans* and a *res extensa*, that is, between something that thinks and something that has spatial extension. As an example, he mentions a triangle, in which necessarily two angles must be equal, even if one would see nothing in the outer reality that confirmed that there is a triangle in the world.[4] It is Descartes, after all, who uses methodical doubt to find a way back to reality after his speculations, by establishing it as an incontrovertible fact that he thinks: *Cogito, ergo sum*.

Pascal's *esprit de géométrie* was taken up by Fontenelle (1657–1757), who lived to be almost a hundred years old. He not only wrote dramatic and lyrical texts but also treatises on geometry and physics. Therefore, he is familiar with the tension between literary intuition and discursive "geometric" thinking. He wants to transfer geometric thinking from geometry to many other areas of knowledge. Because the order, clarity, and accuracy that can be found in good books for some time are due to the "esprit géométrique"[5] which is currently spreading further and can even be found among those who know nothing about geometry. The mutable and unpredictable individual phenomena of poetry and history, characterized by intuition, chance, and unpredictability, are closed to "geometric" thinking. History and its consideration, like poetry, cannot be grasped with the exactness and well-orderedness of geometric and philosophical thinking, which is more suitable for calculating the always same movement of the stars.[6]

In Fontenelle, on the one hand, there are the principles of objectivity and abstraction of scientific and philosophical treatises, and on the other hand, those of intuition, talent, and chance in literary texts. As a synthesis, he proposes the construct of a *poète philosophe*. If in antiquity the intuitive poetic grasp was still in the foreground and the model of the *poeta eruditus* was still present in the Renaissance, then a *poète philosophe* is soon to be expected, in which reason dominates. It is to be hoped that poets will in the future boast more of being philosophers than poets, as reason continues to spread.

The courtier and writer Saint-Evremond (1610–1703) sees things quite differently, regarding the *esprit géométrique* in contradiction to the courtly lifestyle, the *honnêteté*. He finds geometry difficult and advises avoiding it, as it only steals precious time. Arguing from an Epicurean perspective, he says

---

[3] Gusdorf 1974, 233.
[4] Descartes 1960, 60.
[5] Fontenelle 1968a, 34.
[6] Fontenelle 1968a, 35.

it is more in our interest to enjoy the world than to know it.[7] He can only advise serious studies when they deal with things that are related to people, such as moral teaching, which teaches how to deal with passions, politics, which guides proper behavior in the state, or literature, which refines the mind in a pleasant way.[8] Conversation with others perfects the ideal courtier, the *honnête homme*. It is much easier, though inappropriate, to spread one's own fixed area of knowledge before others than to talk about many different trivialities. And what applies to conversation also applies to literary works. It depends on how a man of the world, who wants to spend his free time pleasantly, changes the topics in the interest of entertainment.[9]

In the 18th century, the encyclopedist Jean-Baptiste d'Alembert takes up the paradigm of geometry again. Because the constructs of geometry do not necessarily have a correlate in reality, their level of abstraction is particularly high. For a lexicon, he wishes for short and clear definitions, for which geometry is the model.[10] To avoid circularities, he wants to start with some undefined words, as definitions consist in explanations of individual words by others. He distinguishes sciences that deal with individual facts from those in which discursive reason and reflection dominate. Thus, one has to distinguish natural history from natural science, the history depicted in the Bible from theology. Although the geometer proceeds slowly and step by step in his considerations, he does not have to become alienated from the world, he can also be witty.[11] However, he has to set clear priorities and take into account that polite conversations like those in the salons of the 17th century only distract. Descartes would never have discovered the application of algebra to geometry in the Hôtel de Rambouillet.[12]

In Diderot's view, the zeitgeist has changed and moved away from d'Alembert's position towards the individual circumstances of history and literature. In a letter to Voltaire in 1758, the age of mathematics seems to him to have passed. The general taste has turned towards natural history and literature. Therefore, d'Alembert must adapt and engage in literary activity, which might be difficult for him.[13] The high degree of abstraction of

---

[7] "nous avons plus d'interest à joüir du monde qu'à le connoître." Saint Evremond 1965, 12.

[8] Saint-Evremond 1969a, 12.

[9] "C'est un Homme du Monde, qui dans une grande oisiveté cherche à passer agréablement le tems; qui écrit tantôt sur un sujet, tantôt sur un autre, uniquement pour s'amuser; c'est un Bel-Esprit qui pense à se divertir, et à divertir un certain nombre d'Honnêtes-gens avec qui il est en commerce." Saint-Evremond 1706, préface.

[10] "Les définitions et les démonstrations de géométrie, quand elles sont bien faites, sont une preuve que la brièveté est plus amie qu'ennemie de la clarté." Alembert 1967b, 494.

[11] "C'est un grand géomètre, dit-on, et c'est pourtant un homme d'esprit." Alembert 1967b, 350.

[12] Alembert 1967b, 361.

[13] Diderot 1964, 181, note 1.

geometry appears to Diderot as a disadvantage, as it moves too far away from everyday reality. And if one has to verify and confirm geometric calculations through experience and experiments anyway, then the question arises why not experiment directly, without the detour via geometry.[14] Diderot contrasts the geometer with the genius, who does not proceed discursively, but grasps things intuitively, effortlessly, and without compulsion.[15] In this respect, the genius proves to be the successor of the *honnête homme*. Already Daubenton, who had written the article "Botanique" for Diderot's encyclopedia, had rejected Linnaeus' systematizing method, which, for example, distinguishes animals according to the type of toes and feet, as deductive and proposed precise anatomical and physiological observation as an alternative.[16]

In the dispute between the Swedish naturalist Carl von Linné (1707–1778) and the French naturalist Georges-Louis Leclerc, Comte de Buffon (1707–1788), the former is accused of being too geometric and losing sight of reality, while the latter is criticized for lacking geometric thinking. However, here geometry often stands for systematics. D'Alembert criticizes in his letter of September 21, 1749, to Cramer the lack of geometric thinking in Buffon.[17] On the other hand, Linné was accused of losing sight of the concrete in his classification system of natural order. Buffon confirms this by pointing out the absurdity of genus distinctions based on a single criterion. He refers to Linné's distinctions of plants by size, by leaves, or by the type of reproduction, or of animals by the type of toes and feet.[18] Such classification and taxonomy miss the essence of nature, which takes smaller steps. Rather, Buffon argues, one should consider the whole of the individual object. Its complexity is lost in deductive definitions in the manner of mathematics. Instead, one should observe individual specimens inductively and empirically until lasting impressions are established, from which it is possible to arrive at more general ideas.[19] Buffon accuses Linné of overlooking the diversity of nature when he limits himself to a few formal criteria. It is already a mistake of physics to attribute too much uniformity to nature, a mistake that should not be transferred to botany as Linné does.[20] He turns against scholastic and metaphysical speculations, "des méthodes scholastiques, de grands raisonnements fondés sur une métaphysique puérile ou sur des préjugés."[21]

---

[14] Diderot 1964, 178–179.

[15] Diderot 1964, 20.

[16] Pimenta 2017, 693–711.

[17] Hanks 1966, 27.

[18] "Car pour faire un systeme, un arrangement, en un mot une methode generale, il faut que tout y soit compris; il faut diviser ce tout en differentes classes, partager ces classes en genres, sousdiviser ces genres en especes et tout cela suivant un ordre dans lequel il entre necessairement de l'arbitraire." Buffon 1954, 49.

[19] Buffon 1954, 44.

[20] Buffon 1954, VIII.

[21] Buffon 1954, 27.

Geometric constructions a priori, which explain the universe from a principle, could lead to a unified science, which for Buffon is nothing more than an illusion and a beautiful dream. It is rather individual observations that matter, which should be clearly seen and recognized, so that a literary genius can put them in the right order with spirit and taste.[22] With this, Buffon ties in with Diderot's idea of genius.

When Buffon contrasts his method with Linné's, it becomes clear that knowledge is less tied to the natural phenomena studied, and more to the different methods of observation. Buffon considers it impossible to gain certain knowledge of the first causes and a general order of nature, as the senses always only recognize the effects, not their causes. Therefore, it is not about deriving an artificial and abstract system of plants or animals based on individual organs, but about doing justice to the complexity of nature by describing all parts of a phenomenon in a lively language that individualizes and describes the whole.[23]

Nature appears inexhaustible to Buffon, whereupon he draws on a topos of Pliny the Elder. Therefore, its representation is always unfinished. To capture it, one should make observations without intention and repeatedly, and from the slowly forming lasting impressions create ideas. The illustrations, which present the respective object in all its details or in its habitat, make an important contribution to the visualization of Buffon's descriptions. They also guide the simple and clear style of the descriptions, which in turn draw pictures and should be as instructive as they are interesting.[24]

In Linnaeus' systematizations according to certain characteristics such as "split-footed" versus "single-toed" or according to the number of stamens and pistils, the possibility of a transformation of living beings within a generational line seems impossible.[25] In contrast to this are Buffon's considerations on the monster, that is, a being that does not correspond to the usual characteristics in its respective genus. One might think that the established order of nature is shattered by such a being. Buffon suggests not seeing the monster as an error and exception to the laws of nature, but considering the variations and new compositions against the background of different circumstances as the rule, thereby anticipating the ideas of Darwinism.[26]

Buffon was not only familiar with biology, but also with mathematics and geology. Through experiments, he demonstrated that the earth follows a directed development. It cools down in seven periods from a glowing ball to an ice ball. About 75,000 years have passed since it was torn from the sun. In unpublished writings, he even comes up with millions of years, which

---

[22] "Bien écrire, c'est tout-à-fois bien penser, bien sentir et bien rendre, c'est avoir en même temps de l'esprit, de l'âme et du goût." Buffon 1954, 503.

[23] Bies 2012, 211.

[24] Baere 2007, 614.

[25] Müller-Wille 2011, 112.

[26] Nouailles 2016, 57.

contradicted the church's ideas, which derived about 6,000 years from biblical information. During this development, it was initially completely covered by water, until land surfaces emerged through the lowering of the sea level.[27]

It has been shown that in the tension between systematics and intuition, method, style, and arrangement play a role. While Pascal's *esprit de finesse* refers to everyday observation, the *esprit de géométrie* operates in the realm of definitions and derivations. Descartes had cited the example of the geometric triangle, which, as it is conceivable, cannot be found in reality. Fontenelle praises the clarity and accuracy of geometric thinking, which has already penetrated areas where geometry is not understood. He sees the connection between literature and geometry in the *poète philosophe*, where rational thinking dominates. In contrast, Saint-Evremond holds a different opinion. He believes it is better to enjoy the world than to understand it rationally. His ideal is embodied by the *bel esprit,* who takes pleasure in socializing with distinguished society. The contributor to the great French encyclopedia of the 18th century, d'Alembert, sides with Fontenelle when he demands brevity and accuracy of geometry for his definitions and finds social life in the salons detrimental to knowledge in geometry and algebra. However, this does not prevent him from believing that a geometer can also be eloquent and witty. He also distinguishes the more literary method of natural history, which sticks to the individual, from theoretical natural science. His colleague on the encyclopedia project, Denis Diderot, on the other hand, considers mathematical paradigms to be outdated and prefers the descriptions and experiences of natural history. Another contributor to the encyclopedia, Pierre Daubenton, also criticizes natural science and rejects Linnaeus' systematic method as deductive, while he advocates observations. In fact, Linnaeus had introduced taxonomies and classifications according to certain characteristics in the manner of mathematics. His greatest critic, Buffon, considers it impossible and illusory to create a unified science based on geometric constructions that explain the universe from one principle. He therefore accused Linnaeus of overlooking the diversity of nature, which he himself wanted to reproduce with his descriptions of natural phenomena and illustrations. Particularly interesting against the background of the two different approaches of Linnaeus and Buffon are the attempts to explain deviations from the laws of nature, such as the case of monsters.

4.2   RHETORIC AND LOGIC

Honoré de Balzac (1799–1850) aimed to present a realistic picture of French society of his time through his novels. The complete works, which he summarized under the title *Comédie humaine* and provided with a preface, were

---

[27] Seibold 2009, 2023–2029.

to comprise 137 novels and stories, of which he completed 91. In accordance with the cyclical nature of the work, characters appear in multiple novels, providing insights into different milieus and atmospheres. What Buffon had criticized of Linné, namely explaining the universe from one principle, is exactly what Balzac undertakes with the *unité de composition*, which he bases on the comparison of man and animal, his "comparaison entre l'humanité et l'animalité"[28]. However, Balzac does not start from geometric constructions, but from the comparability of man and animal. The *unité de la composition* was a topic that not only the zoologist Geoffroy Saint-Hilaire (1772–1844) defended against the zoologist Baron de Cuvier (1769–1832) in 1830, but that in different forms already occupied Swedenborg , Saint-Martin, Leibniz, Charles Bonnet, and Needham, as Balzac emphasizes in the preface to the *Comédie humaine* from 1842. Here, a connection between spirituality and science is also indicated, which already characterizes Balzac's earlier novels *La peau de chagrin* (1831) and *Louis Lambert* (1832). If he further asks how best to make a plot with several thousand characters so interesting that he pleases a poet, a philosopher, and the great crowd at the same time with impressive images, then the attempt to connect knowledge and fiction becomes apparent.[29]

Balzac sides with Geoffroy Saint-Hilaire, who saw analogies between human society and the animal kingdom based on the *unité de la composition*. Saint-Hilaire considers the mathematical style of physicists to be of little help, as it cannot capture natural phenomena, and praises Buffon's style, which is criticized by the "géomètres de son temps" because he does not share their analytical approach.[30] He considers it unhelpful to convert every observation into mathematical formulas. Mathematics is only useful for verification, not for discovering new things. Moreover, the mathematized representation only offers what has been put into it beforehand, only in a different language. While this is possible in physics with its simple objects, it is not sensible in physiology with its complex bodies, where there always remains an inexplicable residue, which one might be content to label as vital forces, *forces vitales*. The phenomenon of the *force vitale* does not have the value of a substance for the literary critic Hippolyte Taine, but is a kind of relationship like fate, for example. He proves this by the method of deduction using the example of the digestion process. From the sentence "The animal digests," numerous processes and details can be derived as necessary: transformations by stomach acid, different movements of the stomach, effects of the intestine, etc. Thus, according to Taine, a catalog of details can be compiled. As an alternative method, Taine presents induction, which, starting from numerous phenomena, arrives at the end of all mental and natural sciences at a few concise

---

[28] Balzac 1976, 7.

[29] Klinkert 2020, 213–214, 226.

[30] Saint-Hilaire 1838, 9.

general formulas, even at the *unité de l'univers*.[31] Once in possession of these basic sentences, the concrete can then be derived from the general again. That Taine's method of induction has a certain similarity with the rhetorical method of text extension, the *amplificatio,* becomes clear in Balzac's *Traité de la vie élégante*, where he sees the definition as a summary of a complex reality with numerous individual phenomena: "Définir, c'est abréger."[32] After the definition, he wants to demonstrate the elegant life with its implications. From the development and concretization of a single idea, an entire book is created. If someone has carried a thought summarized in one sentence for a week, he will be able to find an astonishing amount of ideas that have formed unconsciously.[33] Thus, from the small original idea, a large tree with numerous branches is formed.[34] The image that Balzac makes of the deduction of the more concrete from the general is reminiscent of the tree that Porphyry drew to illustrate the Aristotelian logic and doctrine of definition by indicating the next higher genus, the *genus proximum,* and naming the differences to things of the same kind, the *differentia specifica*. In it, the very general is at one end and the concrete details are branched at the other end. Balzac seems to have been familiar with the doctrine of the further subdivision into species and genera. He criticizes it in the author of the *Art de mettre sa cravate:* "Des division, des séparations de genres, des classifications, des prohibitions, toute une législation aristotélique."[35] Thus, a proximity of definition, induction, and deduction in logic to the doctrine of commonplaces, the *loci communes,* in rhetoric becomes apparent. Both serve to generate texts. Marmontel recommends the *loci communes* in the context of his *Leçons d'un père à ses enfants sur la logique,* which was posthumously published in 1802. They are as indispensable for the writer as the colors for the painter, although they are not responsible for whether a Titian or Raphael results from it. This is how he is quoted by Eugène Thionville, who teaches rhetoric at the secondary school of Limoges, in his dissertation *La théorie des lieux communs dans les topiques d'Aristote et les principales modifications qu'elle a subies jusqu'à nos jours*.[36] For Thionville himself, they are a universally applicable method that provides the means to discuss any topic.[37]

---

[31] "Nous découvrons l'unité de l'univers et nous comprenons ce qui la produit." Taine 1888, 368.

[32] Balzac 1938, 56.

[33] "à grouper autour de cette innocente épigramme la multitude d'idées qu'il avait acquises à son insu et qu'il s'étonne de trouver en lui." Balzac 1968, 36.

[34] "Ainsi l'ébauche vécut et devient le point de départ d'une multitude de ramifications morales." Balzac 1968, 36.

[35] Balzac 1938, 49.

[36] Thionville 1855, 121.

[37] "méthode universelle qui doit nous fournir les moyens de raisonner sur quelque sujet que ce soit." Thionville 1855, 129.

For Balzac, natural history is an important model when he wants to present his social groups, *expèces sociales,* just as Buffon has presented his zoological species, *espèces zoologiques.* In the world of humans, the soldier, the worker, the lawyer, the idler, the scientist, and the politician can be distinguished just as in the animal world the wolf, the lion, the donkey, the raven, or the sheep. However, while there is only one type of female in the animal world, the merchant's wife differs from the prince's or the artist's in humans. One can now emphasize the underlying unity of the different types in the description of nature, like Geoffroy Saint-Hilaire, or consider more the differences between the types, like Cuvier. Where Balzac joins the former, he formulates in his preface to the *Comédie humaine* almost pantheistically, there is only one living being. The creator used the same pattern for all organized beings. However, elsewhere he seems to join Cuvier more when he holds social conditions responsible for the fact that there are as many different people as there are species in the animal kingdom.[38]

Theory seems to Balzac more important than literary fiction. This impression at least arises when he assigns a higher value to his *Etudes analytiques,* which he counts among the genre of physiology, in his preface to the *Comédie humaine* written in 1842, than to his novels.[39] He sees himself as an archaeologist of the social world, a classifier of professions, and a registrar of good and evil,[40] explaining that archaeology in the social field is similar to comparative anatomy in the field of nature.[41] Statics become dynamics when the descriptive inventory of the *Etudes analytiques* gives rise to the fictional novels of the *Études des moeurs,* such as Balzac's novel *Les employés* from the *Physiologie de l'employé.*

Balzac contrasts the orientation towards the natural sciences with the self-confidence of literary work, which has its own rules. Gastrosoph Jean Anthelme Brillat-Savarin (1755–1826), who wrote the first text of the new literary genre of *Physiologie* with his *Physiologie du goût* (1826), sees himself as an author who is also a "physicien, chimiste, physiologue".[42] Like phrenology, which infers the way of thinking from the shape of the skull, physiology aims to include physical nature in the description of human behaviors. The term physiology comes from medicine, where it refers to the knowledge of the human organism and its healthy functioning. Transferred to the fashion

---

[38] "Il n'y a qu'un animal. Le créateur ne s'est servi que d'un seul et même patron pour tous les êtres organisés." "La Société ne fait-elle pas de l'homme, suivant les milieux où son action se déploie, autant d'hommes différents qu'il y a de variétés en zoologie?" Balzac 1976, 8; cf. Guillo 2006, 68–69.

[39] Balzac 1976, 19.

[40] "archéologue du mobilier social, le nomenclateur des professions, l'enregistreur du bien et du mal." Balzac 1976, 11.

[41] "L'archéologie est à la nature sociale ce que l'anatomie comparée est à la nature organisée." Balzac 1976, 1125, Anm. 6.

[42] Brillat-Savarin 1880, 9.

of literary physiologies in France, especially in the years 1840 to 1842, it refers to essays that depict and caricature the manifestations and customs of different professions, types, circumstances, or social groups through ridicule, irony, and illustrations. They give the impression of scientific exactness through elements from natural science such as classification, deduction, and exemplary demonstration.[43] Balzac follows Brillat-Savarin when he suggests in his *Physiologie gastronomique* (1830), which consists of two magazine articles, that human types should not be distinguished like Lavater or Gall do by physiognomy, gait, or skull shape, but by palate and eating capacity, so that a scientific subdivision arises from the glutton and good eater to the taster and gourmet.[44] Balzac himself wrote *La physiologie du mariage* (1829) and *La physiologie de l'employé* (1841).

In *Traité de la vie élégante* (1830), Balzac admires Cuvier, who could infer the entire animal from a single bone unearthed and thus classify it. As a paleontologist, Cuvier had listed the anatomical peculiarities of an opossum fossil to an attentive audience in the quarries of Montmartre, even before the entire skeleton was exposed. The way he identifies the type based on individual details, one could similarly infer the entire furnishing from a single chair in the realm of social life. Thus, archaeology can be transferred from anatomy to the social sphere, and Balzac becomes the aforementioned *archéologue du mobilier social*, who introduces the *espèces sociales*. In *La peau de chagrin*, it is Raphael who, like an archaeologist in the antique shop, assembles the ruins of the past into historical contexts. The image of a Roman empress evokes the culture of imperial Rome, Cicero's bust the historiography of Livius.[45] However, while the estate society used to shape the individual, freedom in France has now created the possibility of self-determination. The perspective shifts from the order of living beings to possible modes of operation. Here, Balzac assigns a special role to life energy, the *énergie vitale*. It is a kind of capital that can be spent at different rates depending on the individual's life path. Taming passions and thoughts prolongs life. This is beautifully illustrated in *La peau de chagrin* by Raphael's shagreen, which can fulfill wishes, but with each wish fulfillment becomes smaller, thereby also shrinking Raphael's lifetime. If life appears as energy consumption, then illness is nothing more than an increase from normal expenditure to waste. The fact that this vitalistic perspective lies beyond the reach of the natural sciences is shown in the novel by the biologist, the physicist, and the chemist, whose analyses are limited to the material surface of the shagreen.

What do Balzac's attempts look like to suggest scientific objectivity in his narratives? In the *Comédie humaine*, the illusion of the historicity of the narratives is created by "objectification markers" that create the illusion of

---

[43] Biesbrock 1978, 287–358.

[44] Balzac 1938, 63.

[45] Föcking 2002, 95–99.

verifiability by making circumstances of the epoch appear as causes. However, historical events in Balzac always only have the status of prehistories, they are not part of the main plot.[46] Unlike Zola, who holds inheritance solely responsible for the decay of his characters, Balzac's failing heroes have various disease histories, which are presented so extensively that they become a characteristic of the realism of the *Comédie humaine*. In addition, the disease histories act as causes for actions or the absence of actions. Thus, after his bankruptcy, Birotteau is physically no longer able to endure the excitements and demands of business activity and to make a fresh start. If he withdraws from business life, this is due to his disease history. On the other hand, there are the limits of the striving for objectification. If Balzac sees passion as the essential characteristic of man, he cannot choose the descriptive-taxonomic representation of the natural scientist, but has to further develop a narrative-historiographical one in the sense of Walter Scott.[47]

Perhaps it is rather a literary method that guides Balzac? The doctrine of the *loci communes* in relation to logic has already been mentioned. The rhetorical exercises of *amplificatio* through the use of *loci communes* were part of the 19th-century school curriculum in France.[48] How the *amplificatio*, the expansion of a short initial text, is possible with the help of *loci communes* is demonstrated using the example of *Physiologie de la femme* (1842).[49] When the external physical beauty and its characteristics are presented, differentiated by age stages from child to old age and by education between banal and elite women, it is about wealth, status, temperament, and lifestyle, criteria that have been available since Cicero and Quintilian under the search formulas *fortuna, condicio, animi natura, studia*, and *quid affectet quisque* or *genus, sexus, habitus corporis, aetas, educatio*, and *disciplina*. When the motives, tactics, means, timing, and accompanying circumstances with which a woman can be won are discussed, the argumentation is *a causa, a modo*, and *a facultate*. Finally, when the situation of women in antiquity, in the Orient, and in France is differentiated, this is done according to the *loci a tempore* and *a loco*. The amplification of the text is made possible by the application of the *loci communes*, which belong to the chapter of *inventio*, the finding of material, in rhetoric. The opinion that thinking should precede writing is held by the rhetoric professor of the *Faculté des Lettres de Paris*, Eugène Géruzez, which he expresses in his *Cours de littérature, rédigé d'après le programme pour le baccalauréat* (1842) citing Horace, but also scientists like Newton and Buffon.[50] The naturalist Buffon, on the occasion of his admission to the *Académie française* in 1753, instead of honoring his predecessor, gave a

---

[46] Küpper 1986, 41, 44.

[47] Klinkert 2010, 134.

[48] Molino 1980, 182–183.

[49] Neufville 1842.

[50] Géruzez 1842, 74.

lecture on style in which he honored literature by comparing it to nature. What distinguishes a literary work, he argued, can be found prefigured in nature, where the entirety slowly and silently emerges from the germs of the individual and matures to perfection.[51] The writer should proceed as nature does.

Even when nature is seen as a model, the capabilities of natural science appear limited from Balzac's perspective. This is already evident in epistemological questions about the location of space and time. In *La peau de chagrin*, the question is asked whether in cognition the world comes into the brain of the knower or whether the brain can freely dispose of space and time.[52] In this novel, the admiration of natural science is contrasted with the satire of scientists. While on the one hand the works of Cuvier are admired, the scientists and doctors in the second part of the novel appear ridiculous. Cuvier is characterized not only as a scientist, but as a magician of numbers and as a poet. The doctors who are supposed to heal Raphael, however, find no clear diagnosis. One considers the soul as a vitalist and spiritualist in the sense of van Helmont, the other follows Cabanis as a mechanist considering the organs of the body, and the third, as an eclectic, has doubts. The scientists are equally divided and unsuccessful in their analysis of the piece of leather, with one thinking as a representative of natural history, another as a physicist, and the third as a chemist.[53]

That the limits of medicine are precisely where psychological causes dominate, Balzac makes clear through mysterious diseases, where the exact diagnosis is lacking. Thus, in *La Cousine Bette*, moral deficiencies in Valérie Marneffe lead to physical ones, when hair and teeth fall out, an unpleasant smell emanates from her, and she feels like a piece of wet earth. Medicine errs when it restricts itself to the physical side to cure diseases. Balzac points out the destructive power of jealousy in *Lys dans la vallée* and holds thoughts responsible for deadly diseases in *La Physiologie du mariage*: "Une pensée peut tuer un homme."[54]

Despite this criticism, medicine and biology gained publicity in the 19th century. Three events of the 19th century that are important for the biologization can be highlighted: In 1856 Johann Carl Fühlrott's discovery of parts of a human skeleton in the Neanderthal, whose flat forehead and conical back of the head correspond neither to a skeleton of today's human nor to that of the ape; in 1857 the publication of the *Traité des dégénérescences physiques, intellectuelles et morales de l'espèce humaine* by the doctor Bénédict-Auguste

---

[51] "C'est que chaque ouvrage est un tout, et qu'elle travaille sur un plan éternel dont elle ne s'écarte jamais; elle prépare en silence les germes de ses productions; elle ébauche par un acte unique la forme primitive de tout être vivant; elle la développe, elle la perfectionne par un mouvement continu et dans un temps préscrit." Buffon 1904, 8.

[52] Balzac 1972, 10.

[53] Neffs 1979, 127–142.

[54] Borel 1973, 145–146, 53.

Morel, who sees decadence in human history as conditioned by the inheritance of diseases and claims, following Rousseau, that every civilizational progress results in a weakening of health; and in 1858 Charles Darwin's lecture, which was published a year later under the title *On the Origin of Species by Means of Natural Selection*. Thus, Morel's theory of degeneration and Darwin's theory of evolution stood in opposition, with the choice, in view of the so-called Neanderthal man only named as such in 1864, of feeling like a perfected ape or a degenerated Adam.[55]

In Balzac, such scientific theories are not implemented in pure form. Rather, he mixes them with rhetoric, history, and metaphysics.[56] This is particularly evident in his novels *La peau de chagrin, La recherche de l'absolu, Seraphita,* and *Louis Lambert*. Thus, Lambert appears under the influence of Swedenborg, Mesmer, Lavater, and Gall as a "chemist of the will"[57], who wants to unite the subjects of physics, chemistry, and biology, which are taught separately at universities. When the passionate chemist Balthasar Claes, who once studied with Lavoisier, ruins his fortune and his family in *La recherche de l'absolu* in search of the unity of the elements through excessive expenditures for his laboratory and his experiments, Faustian striving, alchemy, and the hubris of empirical research are equally criticized.[58] A mathematical theosophy is shown in *Séraphita,* where God as the principle of unity forms the origin of multiplicity.

The fact that geography is not considered one of the highly esteemed natural sciences becomes clear when one considers that it is not mentioned in the preface of the *Comédie humaine*. This is also evident in the character of the small official Phellion, who appears in *Les Employés* and *Les Petits Bourgois* and gives courses in geography and history at a girls' school. However, the subject of geography was only just emerging at the beginning of the 19th century. In 1807, the geographical journal *Annales des voyages* was founded, and from 1831, an aggregation as a university final examination was created for prospective teachers in geography and history. After all, geography was considered a subsidiary science of history. Balzac admired Humboldt as a geographer who wanted to combine science and literature in his writings and pursued a similar approach in his descriptions of milieu as Balzac himself, who localized the actions of the novel characters with descriptions of milieu.[59]

In Balzac, it has been shown how elements of natural science, history, vitalism, metaphysics, and an early form of sociology mix. In addition, logic competes with rhetoric.

[55] Föcking 2002, 281–290.

[56] Thiher 2001, 39.

[57] Thiher 2001, 55.

[58] Thiher 2001, 57–58.

[59] Huaulmé 2023, 7–19.

The *unité de composition* was a postulate of the natural sciences as well as spirituality. Balzac built his analogies between the animal world and the human world on this in the sense of Goffroy Saint-Hilaire. The *forces vitales* make life appear as energy consumption in Balzac, where natural science reaches its limits in the face of passions such as jealousy and excessive thirst for knowledge. The *forces vitales* were also the ones on which Taine illustrated his teachings of logical induction and deduction, which reminded of the rhetorical exercise of *amplificatio* through *loci communes*. If a whole book like a large tree with numerous branches emerges from an idea and Balzac mentions the almost Aristotelian subdivisions into species and genera in *Art de mettre sa cravate* on the occasion of tying the tie, this evokes the Porphyrian tree, which logically exemplifies from the most general to the most concrete through definitions with *genus proximum* and *diferentia specifica*. Can one, like Cuvier, conclude from an excavated bone to the whole animal, also conclude from a piece of furniture to the entire "mobilier social"? When animal species are compared with social types, Balzac sometimes emphasizes the underlying unity, other times the diversity. He at least sees himself in his analytical studies as an archaeologist of the social world, as a classifier and registrar, when he differentiates and classifies eating habits according to the criteria of palate and eating capacity. In his novels and physiologies, on the other hand, he transforms analytical statics into fictional dynamics. When the naturalist Buffon gives a lecture on style, in which he compares the writer's work with the activity of nature, and when Balzac, imitating the natural history narratives of biology, makes rhetorical and logical classifications of social species and genera, literature and natural science are in dialogue.

4.3   Tradition and Advancement

In the case of the Spanish representative of the realistic novel Benito Pérez Galdós (1843–1920), literature and natural science are presented as opposites, but the evaluation fluctuates. In a letter from 1889, Galdós admires the doctors, who in his opinion could better grasp the essence of people than the writers.[60] On the other hand, religious ideas such as forgiveness, communion, confession, or martyrdom also play an important role in his realistic novels.[61]

In *Doña Perfecta*, a book referred to as a thesis novel, the positions of the defenders of tradition and the representatives of technical-scientific progress engage in a heated debate, leaving it open on which side the author stands. The spatial structure of the plot already illustrates the opposing worlds. The young engineer Pepe Rey from Madrid travels by train to the provincial town of Orbajosa. He is supposed to marry Rosario there, and also to provide

---

[60] Weiser 2013, 211.
[61] Noel 2010, 151.

technical assistance in the construction of a bridge. The villagers view him skeptically and soon the village priest provokes him by accusing the modern natural sciences of leading to the downfall of feelings and hopes with their laws. They restrict the freedom of the spirit, destroy the beautiful in the arts, deny poetic inspiration, and question the existence of the soul. They only accept numbers and lines. "La ciencia dice que todo es mentira y todo lo quiere poner en guarismos y rayas [...] la inspiración misma de los poetas, mentira."[62] The reply of the engineer Pepe Rey confirms the facts, but unlike the village priest, he sees them in a positive light, welcoming the farewell to silly dreams and rejoicing that humanity is finally waking up, seeing everything with clear eyes and discarding old sentimentalities and hallucinations. Mars has become Moltke, Orpheus has become Verdi, and Vulcan has become Krupp. There is no Parnassus just as there is no Olympus. And there is an Elysium only in the form of the Champs Elysées in Paris.[63] The confrontation ends tragically. Pepe Rey is shot while trying to flee with Rosario; Rosario goes mad as a result of the events. Thus, Pérez Galdós does not always present the two paradigms as irreconcilable.

In the novel *Marianela*, Pablo loves the poor young orphan Nela. Since he is blind, he admires her dreamy views and considers her the most beautiful woman in the world. He only learns that she is actually unattractive when he regains his sight through a medical procedure. Now Pablo admires the beautiful Florentina, so Nela withdraws until she dies of an illness. Thus, scientific progress has led to Pablo's healing and Nela's death. In the novel, the doctor who performs the procedure first appears as a curious student, who, while cleaning a shirt sleeve, recapitulates the names of the muscles and parts of the skeleton underneath, and then as a fully trained surgeon compared to a genius.[64] The achievements of natural science enable the admirable operation of one with the serious consequence that the emotional life of the other is permanently destroyed. If one were to read the novel as evidence of social Darwinism, one would overlook how important the world of illusions and fantasies is to Galdós.

Doctors often appear in the novels of Pérez Galdós, and disease patterns are described in detail and with expertise. Galdós initially depicted the harmful effects of alcohol in his first novel *La desheredada* (1881) in detail, but then deleted it, probably to avoid appearing too much as an imitator of Émile Zola's *L'Assommoir* (1877). In *Fortunata y Jacinta*, Fortunata, Jacinta, and Juanito Santa Cruz are examples of different alcohol dependencies, as is Don

---

[62] Galdós 1972, 55.

[63] Galdós 1972, 56–57.

[64] "Intrépido y sereno, había entrado con su ciencia y su experiencia en el maravilloso recinto cuya construcción es compendio y abreviado resumen de la inmensa arquitectura del Universo. Era preciso hacer frente a los más grandes misterios de la vida, interrogarlos y explorar las causas que impedían a los ojos de un hombre el conocimiento de la realidad visible." Galdós 2016, 103.

Pito in *Ángel Guerra*.[65] Madness is also not uncommon in Galdós' novels. An example is the title character of *La desheredada*, who is told early on that she is of noble origin and will later be elevated to her rightful status, which is not true and contributes to her failure. Galdós was aware of the disagreements among physicians as to whether madness is hereditary or not. In *Lo prohibido*, it is the protagonist José María who attributes his exaggerated fear of illness, his insomnia, his neurasthenia, and his panic attacks to his family history.

Different paradigms are also demonstrated in Galdós' novel *Miau*. The novel was published in 1888, shortly after the definitive third edition of Menéndez y Pelayo's *La ciencia española* was published, where the Spanish contributions not only in the natural sciences, but in all sciences, including theology, mysticism, and poetry, are listed. The work was intended as an answer to the rhetorical question about Spain's contributions, "Que doit-on à l'Espagne?", which Nicolas Masson de Morvilliers posed in the French *Encyclopédie méthodique* (1782–1832), a continuation of Diderot's 18th-century encyclopedia, and particularly suggested the answer "nothing" with regard to the natural sciences. It was religious traditions that had prevented scientific thinking in Spain. Galdós takes up this discussion in his novel. The title "Miau" is not only the sound of a cat, but the abbreviation for "Moralidad, Income tax, Aduanas y Unificación de la deuda", the main ideas of a plan developed over decades as an employee of the Ministry of Finance, with which the protagonist Ramón Villaamil wants to reform his country's economy. However, his passivity and unsystematic approach prevent any success. Unlike him, his grandson Lusito is open-minded and able to systematically organize his experiences as he curiously walks through Madrid, observes everything, looks at shop windows, and does not lose any of the syllables he picks up from other passers-by. Characterized by observation and rational thinking, he appears as a young and hopeful representative of the new Spanish scientific thinking.[66]

In what contexts can traces of Darwin's theory of evolution be found in Pérez Galdós? If we consider his novel technique, it can be noted that with him, unlike with Zola, it is not the objective biological determinism, but the interaction of socially shaped individual worlds of imagination that dominates.[67] The early Galdós could adopt the classification of social groups and their representatives into species and genera from *costumbristas* like Ramón de Mesonero Romanos and Mariano José de Larra, but also from natural scientists like Linnaeus or Georges Cuvier. The will to typify is evident when Galdós speaks in 1867 in the *Revista de Madrid* of the different types of madness, passion, genius, lovers, enviers, scientists, gallants, merchants, disinherited, frivolous and refers to them as "unidades, caracteres, ejemplares".[68] It is

---

[65] Stannard 2015, 94, 104, 108, 115, 119.

[66] Harkema 2019, 280.

[67] Matzat 1993, 144.

[68] Bell 2006, 10.

important that the individual types are not trapped in a fixed hierarchy, but are in constant change. The idea of degeneration and regeneration is compatible with Darwin's theory of evolution, but has a long tradition, for example in antiquity with the cycle theory of Polybius. After the publication of Darwin's *The Origin of Species* (1859) and *The Descent of Man* (1871), the idea of progress and development is associated with Darwin. Articles by or about Ernst Haeckel had explained the theory of evolution to the Spanish public in the 1870s in the *Revista Contemporánea* and the *Revista Europea* explained Herbert Spencer has transferred the biological theory of evolution to the social area and with his reference to the survival of the fittest in society, he represented the economic form of *laissez-faire*. Galdós often traveled to England, from where he not only brought novels, but also scientific literature.

Nevertheless, in Galdós, the tensions are usually not resolved in favor of one side or the other. On the one hand, he attributes better understanding of human nature to physicians than to writers; on the other hand, Nela's decline appears as a side effect of medical progress. When in *Doña Perfecta* natural sciences and technology become antagonists of spirit, beauty, and poetic inspiration, and the pros and cons of both positions are discussed and staged, it remains open as to who should be given preference. Where the backwardness of Spanish natural science is discussed in comparison to other European countries, Galdós introduces a passivity and aimlessness in *Miau* that he contrasts with openness and rational thinking, qualities that favor scientific thinking. Therefore, in Galdós, scientific and technical progress is not only systematically presented as a contrast to existing traditions. Since it is also something new, it makes the previous habits appear as overcome. This turns the systematic contrast into a historical one, in which the idea of progress comes into play. With its historical implications, it has been fought out as a dispute between the representatives of the old and the representatives of the new throughout history, as will be shown in the following.

CHAPTER 5

History

5.1 THE OLD AND THE NEW

Usually, the engagement with the handed down tradition is directed towards the past, while everything new points to the future. Whether one or the other is more important has been a matter of dispute for centuries. This dispute, which was fought in 17th century France under the name *Querelle des Anciens et des Modernes*, shook the exemplary status of the Greco-Roman antiquity and the principle derived from it, the *imitatio* of ancient models, with its attack. After all, antiquity was considered a normative model and an unattainable pattern. The *Querelle* had precursors in the Greco-Roman Atticism-Asianism dispute and in the Ciceronianism dispute in the 15th century. People found Homer exemplary in antiquity, but also the Aristotelian tragedy poetics and not least the moral strength of this time. The Latinist and literary scholar Hippolyte Rigault generalized the contrast in the 19th century and transferred it to two types of cultural behavior. While some orient themselves towards the past, others prefer their own present, whereby the past and present can be associated with different facts.[1] If the age of Louis XIV is considered as the present and ranked higher than the past, then the idea of progress arises. The young Blaise Pascal illustrated this with the chain of successive generations of people, which should be imagined as a single person who accumulates more and more experiences with increasing age, which is why antiquity belongs to the inexperienced youth and does not represent the peak.[2] While the supporters of antiquity accused their opponents of insufficient knowledge and lack of feeling for literary matters, they criticized in particular the customs of the archaic phase of Hellenism, the

[1] Rigault 1856, 1.

[2] Pascal 1908, 136.

© The Author(s), under exclusive license to Springer-Verlag GmbH, DE, part of Springer Nature 2025
C. Strosetzki, *Literature in Dialogue with the Natural Sciences*,
https://doi.org/10.1007/978-3-662-71319-8_5

Homeric world. Thus, Charles Perrault (1628–1703), known for his collection of fairy tales, in his 1687 eulogy *Le siècle de Louis le grand* presented to the Académie française, as well as in his polemic *Parallèles des anciens et des modernes*, denied the Homeric epics any exemplary character compared to the cultural brilliance of the reign of Louis XIV and France's leading position in world history in view of the achievements of natural science, architecture, and painting. He gave his speech at a time when Louis XIV had recovered from an operation, in the Académie française, whose protector was the king himself and who had given it a prestigious home in the Louvre. As a former employee of the deceased Minister Colbert, who had overseen the royal buildings, Perrault could attribute the brilliance of culture to the glorification of the monarchy. In contrast, Boileau, who belonged to the educated bourgeoisie of the *noblesse de robe*, strove for a deeper understanding of the ancient cultural heritage. It has even been claimed that Perrault, serving the absolutist politics like a cultural functionary, preferred the time of Louis XIV to antiquity, while Boileau maintained distance and thus integrity.[3] In any case, Perrault targeted a broader audience in the *Querelle*, which is why not least the advocates of opinion formation for a pluralistic public stood against the representatives of old-style scholarship.[4]

Historically, literature, art, and taste have often been associated with the past, while technology and natural science have been linked to modernity. However, it should not be forgotten that grammar and rhetoric, as well as arithmetic, geometry, and astronomy, were combined in the *artes liberales*, the former in the trivium, the latter in the quadrivium. In the last third of the 17th century, interest in astronomical and medical questions grew in literary salons. The possession of telescopes was widespread. The *Entretiens sur la pluralité des mondes* (1686) by the early Enlightenment thinker Fontenelle (1657–1757) attempted to convey the heliocentric world system to a non-scientifically educated audience in casual conversations.[5] In his *Entretiens*, Fontenelle, as a *galant homme*, recounts a conversation he had with his hostess, a marquise, during a walk in a park in Normandy. From the advantages of day and night, he moves on to the universe and the star system, which he explains according to the latest findings, as well as the Cartesian doctrine of light. Without the presence of a lady, he would have devoted himself to observing the clear starry sky. But instead, he asks whether the night might not be more beautiful than the day, and receives the gallant answer that the beauty of the day can be compared to the beauty of a blonde woman, which is radiant, but the beauty of the night touches the heart like that of a brunette.[6]

---

[3] Fumaroli 2001, 107.

[4] Disselkamp 2010, 173.

[5] Kortum 1966, 1–28.

[6] "La beauté du jour est comme une beauté blonde qui a plus de brillant; mais la beauté de la nuit est une beauté brune qui est plus touchante." Fontenelle 1991, 17.

Fontenelle grew up under the absolutist rule of Louis XIV and came into contact with the Enlightenment of the 18th century. His uncles Pierre Corneille and Thomas Corneille brought him up in a literary milieu.[7] Among his early works are the *Dialogues des Morts*, written in the manner of Lucian. In the dialogue between Montaigne and Socrates, Fontenelle has Montaigne observe that people have always had the same inclinations, completely independent of reason, which is why their follies are always the same. This is for Socrates an argument against the idea that antiquity was more valuable than the present. It is the distance that makes antiquity appear greater. Viewed up close, great figures like Pericles or Aristides would be very similar to contemporaries. The past benefits not least from dissatisfaction with one's own time. In summary, the character of Socrates maintains that although externals such as clothing, appearance, manners, knowledge, and ignorance may change, the essential, the heart of man, is unchangeable.[8] In another dialogue, Descartes sees it no differently when he doubts the recognition of truth, but admits that hopes and even errors also bring joy.[9]

This idea is illustrated by Fontenelle in his *Digression on the Ancients and Moderns*, where he asserts that the trees of all ages are the same, but not those from different climatic regions.[10] Naturally, the conditions in poetry are different from those in the natural sciences. While in poetry, a lively imagination could produce everything important in just a few centuries, this is not possible in physics, medicine, or mathematics. Here, in view of numerous experiments and considerations, patient slowness is required, with progress never coming to an end and physicists and mathematicians always improving.[11] The dialogue between the Greek physician and naturalist Erasistratos (304–250 BC) and the English physician William Harvey, who discovered the circulatory system, is somewhat more skeptical. Harvey praises modern medicine, which thanks to his discovery is better than that of antiquity. Erasistratos points out that more precise knowledge about the body does not necessarily lead to better healing methods. Knowledge of the circulatory system is no more useful than the discovery of a new star in the sky.[12] It is therefore not surprising that Fontenelle sees human happiness not in externalities, but within. In his treatise *Du bonheur*, he advises limiting oneself and coming to terms with oneself.[13]

[7] Krauss 1969, 10.
[8] "Mais le cœur ne change point, et tout l'homme est dans le cœur." Fontenelle 1990, 86.
[9] Fontenelle 1990, 201.
[10] Fontenelle 1991, 414.
[11] Fontenelle 1991, 419.
[12] Fontenelle 1990, 95.
[13] Fontenelle 1989, 216.

On the surface, the *Querelle des Anciens et des Modernes* is known to be about the question of whether the ancient tradition of the Greeks and Romans or contemporary France should be given precedence in terms of language and literature. In fact, a more fundamental development is indicated, in which reason becomes the sole standard. With the rejection of the ancient models that were to be emulated, previous authorities are sacrificed to the idea of progress and novelty. With the removal of the ancient epics and tragedies, a critical judgment arises that now decides on matters of politics and religion. Even if it initially wants to replace ancient mythology with Christianity, it ultimately subjects the latter to criticism and leads to enlightenment with a reason-guided attitude. In the 19th century, the polarity of authority and criticism of reason, of tradition and modernity, becomes a much-discussed and central constant. Here, Christian elements of Romanticism stand against the positivist features of Naturalism with its model of successful natural sciences.

After the French Revolution, one had to choose between a reactionary attitude and the endorsement of the upheaval. On the side of the *Anciens* are now the defenders of the *Ancien Régime,* who oppose the innovators and the innovations. This makes the debate interdisciplinary. If positioning in the political debate has consequences for how history is viewed, philosophical orientations determine literary trends, and turning towards tradition means turning away from the future, then an intertwining of different disciplines becomes apparent, which can only be understood in their relations to each other. If an author feels committed to tradition, this has consequences for the choice of authors he aligns himself with, and also for the design of the fields of knowledge in which he operates. The same applies to the representatives of the innovators. Thus, there is a sense of belonging in the group of authors on each side, also in the different disciplines in which an author operates. In the period after the French Revolution, it seems even more urgent to align oneself with one of the two camps than during the Enlightenment.

In the first half of the 19th century, spiritualists like Royer-Collard and Maine de Brian stand against materialistic ideologues like Destutt de Tracy and Cabanis, who see ideas only as products of physiological and psychic organisms. There are also opposing attitudes in the political debate, when liberals like Paul-Louis Courier, Saint-Simon, and Fourier distance themselves from traditionalists like Bonald and de Maistre. While Joseph de Maistre (1753–1821), as a monarchist and counter-enlightener, sees the atrocities of the Revolution and the evil in the world as consequences of original sin or as punishments in his dialogues of the *Soirées de Saint-Pétersbourg* (1821), for Saint-Simon the principle of production and for Fourier a future society based on the free development of passions are central. In political theory, La Mennais opposes a Christian-influenced socialism to the socialist teachings of the Saint-Simonists, the Fourierists, and the Proudhonists.

Had Madame de Staël proposed German Romanticism as a model, for the Romantic Chateaubriand, the harmony of the world and the wonders of nature are evidence of Christianity, which produced the art of Gothic cathedrals and promotes and preserves culture and civilization. Several Romantics had collaborated on the magazines *La Muse française* and *Globe* and met in the *Salon de l'Arsenal*. Meanwhile, the spiritualism of a Victor Cousin collides with the positivism of an Auguste Comte, who wants to replace the theological and metaphysical age with a positivist one, in which the natural sciences form the basis.

In historiography, which became a literary genre again in the 19th century, there are on the one hand those who emphasize the ideas behind the historical facts, like Guizot, who wants to confirm his theories through facts, the Romantic Edgar Quinet influenced by Herder, Alexis de Tocqueville, who is looking for reasons for the end of the Ancien Régime, and Louis Blanc. On the other hand, there are those who limit themselves to uncommented facts, like the liberal Augustin Thierry and his successors Barante, Thiers, and Mignet. Both tendencies are combined in Jules Michelet, for whom historiography is an artistic revival of past epochs. Fustel de Coulanges dealt with ancient history in *La Cité antique* (1864), considering the ancient beliefs to be the key to understanding the institutions.

Also in literary criticism, which advanced to a new literary genre, representatives of classical tradition like Nisard face those of a new aesthetic opposed. Taine is responsible for the derivation of literary works from the three factors *race*, *milieu*, and *moment*, which become central for Zola. While Hippolyte Taine practices literary criticism with a positivist determinism, Villemain takes a mediating position between classicism and romanticism and paves the way for the literary critic Charles-Augustin Sainte-Beuve (1804–1869), who wants to analyze and present each writer in terms of origin, character, knowledge, and life circumstances. Against the spiritualist and theological schools of Bonald, de Maistre, and Cousin, authors such as Saint-Simon, Leroux, Reynaud, and Comte represent progress. For them the past is finished and the revolution is the prelude to a new era, in which social conditions are improved and humanity is perfected on the basis of positive scientific facts, which ultimately leads to Comte's cult of the "Grand-Être", which for him is humanity.

Can the opposing groups of the 19th century be compared with the multiculturalist, liberal *Anywheres* of the publicist David Goddhart, who in the 20th century are qualified, mobile, and open to the world, to be distinguished from the sedentary *Somewheres*, craftsmen, skilled workers, and service providers, who face social decline due to global trade and artificial intelligence?[14] Is there even a historical constant emerging here? A psychological study[15] distinguishes "defenders" and "explorers", with the former defining belonging according to ethnic and religious homogeneity and being

[14] Goodhart 2020, 78.
[15] Back et al. 2022.

inclined towards strong leadership personalities, the latter being younger, urban, educated, trusting political institutions, and thinking pluralistically. Although the terminology used here is as evaluative as it is reductionist, it nevertheless shows the *Querelle* as a model that is still referred to even after the early modern period. In the early modern period, at least, the dispute between advocates of a flourishing cultural tradition and defenders of progress, often defined in a technically scientific way, was still undecided.

5.2   Cycles

While in the *Querelle* a clear distinction is made between past and future, thus thinking linearly, this contrast is lost in cyclical thinking, where circular movements form the model, allowing decadence to follow bloom and bloom to follow decadence. The nationalist cultural historian Oswald Spengler (1880–1936) opposes a flat optimism of Darwinian imprint with his work *Der Untergang des Abendlandes. Umrisse einer Morphologie der Weltgeschichte* (1918). For him, world history is an amorphous stream in which high cultures stand out as individual zones of densified interaction. High cultures are temporally and spatially limited, with the boundaries not coinciding with the respective religious, linguistic, and political ones. Given the importance of boundaries, it is understandable that Spengler gives priority to foreign policy over domestic policy. Where domestic policy dominates and pursues its own, usually materialistic goals, decay, the falling out of form, begins.[16] Decadence can thus also be interpreted as a consequence of growth, such that a simple life under difficult circumstances encourages achievement, whereas power, when linked with wealth and luxury, leads to slackening. Nietzsche argues in the same direction and criticizes the lack of the will to power as decadence. While he rejected weakness and lack of vitality as decadence, in contrast to him, Paul Verlaine along with the authors of the magazine *Le Décadent* (1885–89) affirmed decadence as refinement.

The eulogists of the past, the *laudatores temporis acti,* were already numerous in antiquity. But this fact raises the question of whether decadence exists at all, since its presence was lamented in all times. "Decadence would then be the reproach of the ill-tempered and fearful, who always only recognize decay in every change."[17] On the other hand, describing signs of cultural depression and criticizing them can certainly contribute to the positive development of society.

The cyclical thinking of antiquity could start from the empirically known rhythm of day and night or from the recurrence of the seasons and transfer these familiar cycles to larger historical periods. The famous myth of Hesiod of the four ages, the golden, silver, bronze, and iron, is based on the idea

---

[16] Simson 2007, 732, 739.

[17] Bohrer 2007, 657.

of decay, followed by bloom. Specifically, Aeneas in Virgil provides an example of the latter when he recognizes in the underworld in his famous successor Augustus the one who brings back the golden age of Rome.[18] In Virgil, Augustus is evaluated positively, unlike in Tacitus. In the 15th book of his Metamorphoses, Ovid evokes a cycle theory in which he compares the cycle of years with the ages of humans, so that spring is assigned to the boy, summer to the young man, autumn to mature age, and winter to the old man. Similarly, one can see how some peoples grow and others decay. The older Seneca divides the history of Rome according to the ages of a human: The time of Romulus corresponds to infancy, the rest of the royal time to boyhood, the time until the end of the struggle with Carthage to youth, the time of world domination until the beginning of the civil wars to manhood, and finally the monarchy with the loss of freedom corresponds to old age, in which Rome cannot maintain itself.[19]

In ancient Greece, Anaximander and the Pythagoreans metaphorically speak of a "great year" and mean by it an age of the world, after the end of which the same returns. Plato sees an analogy between the development of plants, animals, humans, and state structures. The ancient Greek historian Polybius (200–120 BC) is the first to apply the cyclical model to concrete historical processes in the sixth book of his history of Rome. He starts from the theory of the "great year", according to which before the beginning of a new state the old one finds its end through natural disasters. At the same time, he distinguishes between internal causes of decay and external occasions, which are of a random nature. But the internal causes lie in the circularity of being, that is, in the fact that everything after a rise to greatness undergoes a descent to decay. Thus, the invasion of the barbarians into the Roman Empire appears only as a reinforcing occasion compared to the necessity of the descent after the bloom has been reached.

Even Plato saw his time as a decline. It was preceded by happier times, accompanied by God, which have now passed, as humanity has been left to its own devices, as shown in the transition from aristocracy to ochlocracy and tyranny.[20] This cyclical model is replaced by an eschatological one in early Christianity.[21] Now, the concepts of paradise and expulsion from paradise due to original sin, redemption through the crucifixion, and the Last Judgment dominate. Against this backdrop, the fall of the Roman Empire could be seen as a prerequisite for the onset of immediate end-time expectations. However, Augustine explained that this was the wrong approach, considering the Roman Empire not as fallen, but as having transitioned into the Frankish Empire in the sense of a *translatio imperii* idea. Reformation

---

[18] Schlobach 1980, 40, 48.

[19] Klingner 1965, 495; see also Häussler 1964, 313–341.

[20] Widmer 1976, 838–846.

[21] Löwith 1961.

movements of the late Middle Ages accuse the Church of increasing signs of decay and lament its departure from early Christianity. Augustine, along with Cato Minor, points out that the growth of external goods leads to the gradual loss of internal values.

In the Renaissance, humanists could use the cycle theory to position themselves after the Middle Ages and the rediscovery of antiquity in the ascending movement of a cycle with peak expectations. Now, the time between the fall of the Roman Empire and the Renaissance could be considered a low point, from which a new peak is targeted.[22] The emerging humanism leaves the eschatological component unconsidered and, with its admiration for antiquity, can reject everything that led to its end. Thus, it laments the end of the Roman Empire as a decline of ancient culture. Edward Gibbon in 1776, in his *History of the Decline and Fall of the Roman Empire,* attributed the decline of the Roman Empire to Christianity with its ideas of charity[23], although on the other hand he asks the astonished question of why the Roman Empire was able to maintain itself for so long. The Romantic Novalis later opposed Gibbon's regretful attitude towards the decline of the Roman Empire with a supportive stance. Since all later nations were built on the ruins of Rome, Rome should be attributed a sacrificial role and a continuation in another form should be welcomed. The idea of a decline not as a total end, but as the termination of a form of appearance, whose substance remains and dialectically finds a continuation in another form, would find further followers in Georg Wilhelm Friedrich Hegel and Jacob Burckhardt.

As is well known, Machiavelli also dealt with Roman history. In his *Discorsi* on Titus Livius, unlike Tacitus, he places the Roman Republic at the center. The Roman power state had to constantly expand by military means to maintain itself. Since states, like people, age, Machiavelli presents a cycle of states in which the state, whose life force has expired, falls prey to a neighboring state. State reason is defined as the knowledge of the means suitable for maintaining and expanding a state.

In the 18th century, Voltaire distinguishes in the introduction to his portrayal of the age of Louis XIV four particularly happy ages: that of Pericles in Greece, that of Caesar and Augustus, associated with the names of Lucretius, Cicero, Livius, Virgil, Horace, Ovid, Varro, and Vitruvius, the Florence of the Medici, and the century of Louis XIV. The criterion for a golden age is therefore not state or moral strength, but culture. Quite differently, the Enlightenment state theorist Montesquieu (1689–1755) sees it in his work *Considérations sur les causes de la grandeur des Romains et de leur décadence* (1749). For him, the Roman imperial era, which Voltaire describes as a golden age, is already marked by decadence. Although not a cultural golden

---

[22] Schlobach 1980, 28–30.

[23] Myers 1989–90, 223–238.

age, he recognizes in the early days the great period of Rome's rise, when Romulus and his successors were constantly embroiled in wars.

An entanglement between rise and decay was already achieved in the *Querelle des Anciens et des Modernes*, by placing the dwarf on the shoulders of the giant, who sees further than the giant, even though he is smaller. Thus, modern authors may be smaller than the ancients, but they know more. In the Enlightenment, the cyclical epoch metaphor is often modified and short-ened under the increasing influence of the idea of progress, in that the phase of decadence is omitted and one sees oneself at the beginning or on the way to the peak.[24] The talk is now often of the growth of culture, of the rising sun and increasing light of knowledge. In the course of the year, it is no longer spring or summer that appears as the peak, but autumn, as this is the time of harvest. Thus, it is no longer about the blossom, but about the fruit, i.e., the product. Especially in the second half of the 18th century, before the start of the revolution, one sees oneself according to the cycle theory, similar to the Renaissance, before a new beginning, which was preceded by decay. Such a decay could be found by Montesquieu in the Roman imperial period as well as in the absolute monarchy of his present.

Perrault was the one who, in the *Querelle des Anciens et des Modernes* as a representative of the Moderns, gave the consciousness of decadence a cul-tural-historical dimension by describing the art of the time of Louis XIV as a peak, after which there would only be *décadence, corruption,* and *désordre* for example in oratory, but also in human nature. If one now considers that in the ancient conception of the hitherto unidentified and therefore Pseudo-Longin named author moral decay and the decline of the republic are linked, then Perrault's position can be seen as a preparation for that public debate about political decadence in the first decades of the 18th century, in which Montesquieu also intervened with his *Considérations*, whose time-critical ori-entation and topicality were quite apparent to the contemporary reader. The decadence of Rome in particular makes it clear why the Enlightenment think-ers could also fear experiencing a relapse into barbarism like the Romans. Now it is Voltaire who, with his *Siècle de Louis XIV*, which had been in pro-gress since 1732 and was published in 1751, could let the admiration of the Sun King turn into a feeling of decadence of the present. He speaks of the advances of bad taste, which the foreigner could observe in Paris after the time of Louis XIV. The mediocrity of literary works appears as a sign of deca-dence. In essence, Rousseau only continues Voltaire's criticism of the state of decay of the present, by setting an original "golden" age as the age of great-ness, not the preceding century, and noting a decay brought about by arts, sciences, and luxury not in the arts and sciences, but in man himself. The Rousseauian idea of decline was to find its continuation in the medieval admi-ration of Romanticism.

---

[24] Schlobach 1980, 305–307.

It is noticeable that the criteria for the flowering and decay of an era are not natural scientific ones. As an exception, one could perhaps mention Spengler, according to whom decadence begins where domestic politics, materialistic goals, and luxury dominate. The circularity of being determines a dynamic of its own and the vitality of a state appears as the actual cause for rise and fall, against which other events are only external occasions. Criteria for a golden age are political and cultural greatness. If the Roman Empire is admired, it is either the political expansion or the cultural bloom with Lucretius, Cicero, Livy, Virgil, Horace, Ovid, Varro, and Vitruvius. If Voltaire sees the time of Louis XIV as a cultural peak, then it is because after that there were only mediocre literary works. If Christianity builds on the ruins of Rome in the sense of a *translatio imperii*, then political, religious, and cultural standards apply. Natural sciences seem largely irrelevant for judging the bloom or decadence of an era.

5.3   Origins, Beginnings

If one wants to evaluate one's own era, a comparison with the past is useful. In the *Querelle des Anciens et des Modernes*, it was compared with a cultural heyday in the past. It is even more radical to compare it with the beginnings of human history. People in the pre-social state of nature were imagined to be wild, irrational, and uncivilized. Therefore, they do not possess the achievements of the natural sciences nor literary education. The French words *salvage* and *sauvage* and the Spanish *salvaje* are derived from the Latin *silva*, meaning forest, which gives rise to the association of the wild with the place of the forest. In the Middle Ages, the concept of a wild man was common, who is driven more by instinct than by his will, to whom politeness is unknown, and whose behavior is as rude as his appearance. Thus, in the works of the founder of the genre of courtly romance Chrétien de Troyes (1140–1190), Yvain, who wanders through Brocéliande Forest accompanied by a lion, answers the question of what lies behind his impetuous figure: He is a man,[25] as Thomas Aquinas in reference to Aristotle emphasized the necessity of human sociability and rejected the *homo silvestris*.[26]

In antiquity, Homer's *Odyssey* can be cited, in which the Cyclopes and Polyphemus have characteristics of the wild man. Hercules, dressed in the lion's skin of Nemea and of superhuman strength, appears no less like a wild

---

[25] Bernheimer 1952, 5.

[26] "Dico autem bestiales, puta esse hominem, quem dicunt praegnantes recidentem pueros devorare, vel qualiter bestialitate gaudere aiunt quosdam silvestrium circa pontum." S. Thomae Aquinatis, In decem libros ethicorum Aristotelis ad Nicomachum expositio, Rome, Marietti, 1949, p. 368; But since it is destined for man to live sociably, because he cannot suffice for the necessities of life if he remains isolated, so the society of many must be so much more perfect, the more it is able to correspond to the needs of life. Thomas Aquinas 1971, 9; "homo haturaliter est animal politicum vel sociale." Thomae Aquinatis 1881, 372.

man. Horace portrays the first humans as particularly combative in his Satires. He reads their civilizational progress in the refinement of weapon technology. While initially they hit with fists, then scratched with nails, later they used sticks and then invented weapons.[27] At the end of the state of war, the introduction of language, city building, and legislation follows. In the 18th century, the zoologist Carl von Linnaeus classifies the *homines silvatici* or *homines silvestres*, children abandoned in the forest who are raised by wolves, bears, chamois, etc. Linnaeus distinguishes, among others, the *Homo Americanus, Europaeus* as well as the *Homo ferus* and mentions among several examples the *Juvenis lupinus Hessensis,* a Hessian wolf boy, the *Juvenis ovinus Hibernus,* an Irish sheep boy, and the *Juvenis bovinus Bambergensis Camerarius,* a Bamberg cattle boy.[28]

That the wild man lacks not only reason but also the ability to speak is evidenced by Shakespeare's character of the wild Caliban in the play *The Tempest,* first performed in 1611. Prospero, who lands on a deserted island, finds him there and makes him his servant. A prominent revival of the theme is the novel *Robinson Crusoe* by Daniel Defoe (1660–1731), which was published in London in 1719. The title character already possesses civilizational knowledge when, as a castaway on an island, he sees the originality and authenticity of nature. Only gradually does he change it with tools and objects that he rescues from the shipwreck, thus opposing nature with his civilization. Robinson episodes can already be found in the 9th century in the stories *One Thousand and One Nights,* in *Sindbad the Sailor's First Voyage,* in the 13th century in the second *Aventiure* of Gudrun, and in the 16th century in the 67th novella of the *Heptameron* of Queen Margaret of Navarre.[29]

The literary genre of the *Bildungsroman* or educational novel originates from the original state of human nature as a *tabula rasa* and demonstrates how this natural state is overcome through education and upbringing. The nature of this initial state often has quite idyllic and Arcadian aspects. In Gracián's *Criticón,* Andrenio is raised by animals and first discovers the beauty of nature before he embarks on his allegorical journey with Critilo, at the end of which he is rich in experiences. Here, the insights gained are in the foreground, not civilizational achievements as in *Robinson Crusoe.*

A precursor to the *Criticón* is the *Bildungsroman* titled *Hayy ibn Yakzan,* written in the 12th century by the Arabic philosopher Ibn Tufail, which refers to the protagonist of the same name. He is abandoned as an infant at sea, lands on an island, and is raised there by a gazelle. In this natural state, he now has to develop and educate himself consistently and autodidactically, in which one can discover the Aristotelian idea of education, according to which the goal of the perfect personality is already laid out in nature as entelechy.

---

[27] Horatius Flaccus 1953, 26.

[28] Bien 1971, 287.

[29] Bien 1971, 279.

He goes through all stages of education, unfolds his physical life force, trains his sensory perception, practices technical skills, develops discipline in logical thinking, until he gradually advances from the realm of physics to that of metaphysics and begins to touch the mysteries of life and existence.[30] The author Ibn Tufail lived in Arabic Andalusia. He was vizier and personal physician to the Caliph Yusuf, with whom he also introduced the younger Avicenna, who in turn had written an allegorical recitation with the same title. In it, however, the human soul is seduced by the demonic and evil, thus introducing a Neoplatonic dualism of matter and idea, sensuality and reason. While in Avicenna the higher knowledge occurs in an enlightenment of the human spirit through the ideas in the sense of Plato, in Ibn Tufail, intellectual knowledge is based on sensory perception. Ibn Tufail's novel was published in 1671 in the original Arabic with a Latin translation by Edward Pocock in England under the title *Philosophus autodidactus, sive epistola Abi Jaafar Ebn Tophail de Hai Ebn Yokdhan, in qua ostenditur quomodo ex inferiorum contemplatione ad superiorum notitiam ratio humana ascendere possit.*

The natural state of humans in the 16th century was far more critically assessed by Bodin and Vives. Both advocated for its rapid overcoming. In his 1566 Parisian publication *Methodus ad facilem historiarum cognitionem*, the skeptic Jean Bodin assumes an original state of animal wildness, in which humans scattered in the forest and field, like wild animals, only possessing what they had violently seized. Progress only came about through inventions, arts, and sciences.[31] According to Juan Luis Vives, although man is created for community, he is harsh towards others due to his self-love, claiming as much as he can for himself.[32] Vives credits the introduction of justice for overcoming this situation in *De causis corruptarum artium* (1531), which curbs greed and prevents injustice. This suggests a position that desires the overcoming of a natural state characterized by antagonistic interests through the introduction of socially guaranteed justice. Marin Mersenne (1588–1648), as a theologian, dealt with the original state of man and wrote a lengthy chapter on Adam in his Genesis commentary, who before the fall of sin had already carried the seeds of all known sciences within him.[33] The theologian and natural scientist Pierre Gassendi (1592–1655) refers to Hobbes in his ethical writings and adopts his characterization of the state of nature.[34] Not only in philosophy but also in theology, the natural state is a much-discussed topic. Initially, it was generally about the question of the existence or

---

[30] Behler 1965, 353.

[31] Zöckler 1879, 113–114.

[32] Vives 1990, 553–555.

[33] Zöckler 1879, 18; already the church fathers and after them the scholastic dogmatics of the 17th century had attributed a triple perfection to the man not yet fallen through original sin, i.e., the man in the original state: an intellectual, ethical, and aesthetic one. Zöckler 1879, 11, 24. On the topic of progress and perfection versus decadence in this context: Strosetzki 2008.

[34] Ludwig 2005, 18.

possibility of a *status hominis in puris naturalibus* or a *status pure natura-lis*. In the confessional disputes of the 16th century, Suarez represented the Catholic position by quoting colleagues who had considered a purely natural state (*pure naturalis*) that, although it has never actually existed, can still be thought of as possible.[35] For the Protestant opposition, Quenstedt formulated in 1685: "The Scholastics and Papists add a *status hominis in puris natu-ralibus*; but man has never been in such a state, and he could not be."[36] Both Christian positions agree to deny the existence of a state of nature, but they diverge regarding its possibility. This possibility becomes conceivable if, like the German philosopher Samuel Pufendorf[37], one declares man as a substance to which moral, cultural, and social qualities are merely added as accidents. However, if one decides with Aristotle for the idea that the truly natural state of man as a *zoon politikon* is a social one, then civilization is part of human nature.

The most prominent representative of this negative evaluation of the state of nature is Hobbes, whose influence—not least because he was in exile in Paris between 1640 and 1651—should not be underestimated in France. He had befriended there with the Franciscan Marin Mersenne and with Pierre Gassendi, who celebrated the publication of an excerpt of his main work *De Cive* in Latin in 1642 with enthusiastic congratulations. Hobbes argues that the state of humans outside of civil society, the natural state, is only the war of all against all, and that in this war everyone has a right to everything. This would imply that all humans want to get out of this miserable and abhorrent state as soon as they realize its misery. However, this is only possible if they renounce their right to everything through contracts.[38] So, for Hobbes, it is not about the morally good life, but about bare survival. His doctrine of the natural state can thus be considered as the anthropological basis of his political theory. The insight into natural insecurity establishes the equality of humans, which is initially anthropological and not political. Hobbes' political philosophy begins with the doctrine of the natural state, in which there are neither state and state institutions nor subjects or rulers. Humans are like mushrooms sprung from the earth, with no one responsible for the other. However, this is not to be imagined as a paradise or a golden age, but as a state of constantly threatening violent death, in which life is lonely, miserable, and short. Ambition, competition, and scarcity of goods ensure that everyone is a wolf for the other.[39] Hobbes ties in with the teachings of Christian natural law, but sees the basis of legitimacy of the state not in the divine will, but solely in the will of the individual citizens.[40]

---

[35] Bien 1971, 296.

[36] Bien 1971, 296.

[37] Pufendorf 1711; Pufendorf 1712.

[38] Hobbes 1966, 69–70.

[39] Hobbes 1996, 104.

[40] Ludwig 2000, 97–99.

In a secularized form, the discussion about the natural state is continued by the French Enlighteners of the 18th century. For Montesquieu, nature can be the initial, which is revealed by disregarding everything institutional and historically evolved: "Pour les [lois de la nature] connaître bien, il faut considérer un homme avant l'établissement des sociétés."[41] As with Suarez, the *status naturalis* is not assumed to be factually historical, but is understood as a purely hypothetical state that must be assumed for methodological reasons. The Enlightenment also secularizes theological positions in another respect. Comparable to late scholasticism, which understood the natural state as the complete self-reference of the individual, Rousseau formulates in *Émile*: "L'homme naturel est tout pour lui; il est l'unité numérique, l'entier absolu, qui n'a de rapport qu'à lui-même."[42] From this isolation, Rousseau concludes that humans in the natural state lack language and the ability to reflect. Rousseau's natural humans are not sociable, as they do not need the community.

When Rousseau answers the prize question of the Academy of Dijon, whether the restoration of the sciences and the arts has contributed to purifying the morals, he admits that manners are less rustic than they used to be, and taste is refined. However, he is concerned with moral customs, which are responsible for the happiness of humanity. He does not reject progress in principle, only when its negative effects disregard the common good and the *philosophes* let themselves be guided by missionary zeal and intolerance in their rule of opinion. Sciences and technology have indeed brought about certain vices and evils, but they could also serve to contain them.[43]

That Rousseau does not view the original state of man positively is evidenced by his stance on the phenomenon of the orangutan. Edward Tyson had published the work *Orang-outang, sive Homo sylvestris: Or the Anatomy of a Pygmie Compared with That of a Monkey, an Ape and a Man* in 1699, in which he classified the orangutan in the hierarchical order of man and animal as an intermediate link—an assessment that in the 18th century Linné and Maupertuis agreed with. Rousseau, on the other hand, argues that the wild humans mentioned in travel reports are probably people who are still in the initial natural state and live scattered in the forests, where they would not have had the opportunity to develop the ability of language.[44]

With his discussion of the advantages and disadvantages of the sciences, Rousseau is not alone. In the Weimar Classic, Johann Gottfried Herder

[41] Montesquieu 1973, 98.
[42] Rousseau 1964, 9.
[43] Rehm 2015, 47–60.
[44] Lettow 2015, 95.

(1744–1803) repeatedly dealt with Rousseau between 1775 and 1781 and, in contrast to him, claimed that the sciences have always been part of human societies and cannot be banished from them. Their effects depend on the correct use, which is only possible in a free government like Athens and not in patriarchal or despotic forms of rule. When he compares the exemplary culture of the ancient republics with the moral decay and despotism of his time, he relies on David Hume's *On the Rise and Progress of the Arts and Sciences* (1742) and Adam Ferguson's *An Essay on the History of Civil Society* (1767).[45]

The German Enlightenment philosopher Immanuel Kant (1724–1804) also develops a contrast starting from the beginning of human history. In *Mutmaßlicher Anfang der Menschengeschichte* (1786), he starts from the Old Testament (Genesis: 1. Moses 2–6) and takes up the thesis of the conflict of culture with human nature. For him, nature begins with the good because it was created by God. But then begins the history of human freedom, which introduces evil as the work of man.[46]

Three different meanings of the term "natural state" can thus be distinguished. It can mean chronologically the primordial state of man, politically an original situation in which there was no government power and no state, or finally, in terms of civilization, a stage in which the arts and sciences were least advanced. Rousseau now calls on scientific academies to send expeditions with philosophical observers to underdeveloped countries, who, based on their observations upon their return, write a history of humanity from the beginning, which helps it to understand itself. He assumes that the original man had not yet developed moral standards, but—here he distinguishes himself from Hobbes—was naturally good-natured.[47] Rousseau may have known Turgot's 1750 speech at the Sorbonne, *Tableau philosophique des progrès successifs de l'esprit humain*, in which the technical-scientific development is considered the starting point for the moral development of man. He is of the opposite opinion on this point, as he considers it a source of decline, inequality, and particular interests. However, Rousseau does not go as far as the cynics of antiquity, who saw the natural life of animals as a model for man in their criticism of civilization. Rather, he envisages a rural idyll. Vivid examples of this are his novels *Émile,* where the young growing man should keep himself free from egoism, greed, and luxury, and *Nouvelle Héloise,* where the negative influences of the competitive society are kept away in a self-sufficient household.[48] Technical progress promotes luxury and convenience, and undermines military as well as moral virtues, by weakening judgement. Natural sciences, according to Rousseau, bring about those techniques that weaken the social competencies

---

[45] Zurbuchen 2015, 200–201.
[46] Zurbuchen 2015, 208–209.
[47] Lovejoy 2012, 492–496.
[48] Müller 2013, 39, 43, 47.

of humans.[49] The mythical figure of Prometheus, who had stolen fire from the gods and brought it to humans, appears in Rousseau as the originator of harmful civilization, which not only weakens virtues, but in turn has its origin in human vices. Thus, superstition gave rise to astronomy, ambition to rhetoric, greed to geometry, and vain curiosity to physics.[50]

It can therefore be concluded that in the origins of man, the absence of language and sociability is initially observed. It is disputed whether animal savagery and war of all against all or natural kindness and sociability are characteristic of the beginnings. When the wild man is clothed in furs, this is an indication of the lack of a technique for making clothing. And when Horace measures civilization by the degree of refinement of weapon technology, then technology and natural sciences also seem to contribute to overcoming the state of nature. Finally, the Arabic educational novel, which initially conceives human nature as a *tabula rasa*, traces a development from technology through logic to metaphysics and physics. For Herder, the sciences have always been part of human existence, while for Turgot it is the technical-scientific development to which man owes his moral advancement, whereas for Rousseau it produces inequality and weakens social competence.

5.4   PROGRESS

Most French Enlightenment thinkers were committed to progress. However, they saw this primarily in the new thinking models in England, which were oriented towards the scientific approach. Voltaire's *Lettres anglaises (or Lettres philosophiques)* are not only a plea for the parliamentary English state system, which appears superior to the French monarchy, but also for English philosophy. He refers to Bacon as the father of experimental physics and praises Locke's precision and systematic approach with the words "Jamais il ne fut peut être un esprit plus sage, plus méthodique, un Logicien plus exact que M. Locke":[51] while Descartes appears to be overtaken by Newton. That Voltaire prefers the English model to the French is not only evident in his *Lettres philosophiques* (1734), but also in his *Élements de la philosophie de Newton* (1738). In the *Élements*, he first addresses metaphysical questions and then presents Newton's optics and mechanics in detail. The fact that he paid so much attention to an author like Newton, who was considered heretical in France, and praised the worldly and action-oriented English way of life, but also Locke and Shakespeare, and saw England as a country of philosophers, distanced him from the French way of thinking of a Descartes or Pascal.[52] Already in the 15th letter of his *Lettres philosophiques* he dealt with Newton's

---

[49] Fischer 1996, 293.
[50] Klinkert 2010, 61.
[51] Voltaire 1964, 57, 61.
[52] Calderón 2015, 30.

transfer of gravitational forces from the earth to the planets, sun, and moon. However, this leads him with Newton to the idea that matter could only have received gravity from God.[53] In the writings in which Voltaire deals with Newton, he leaves the mathematical aspects untouched. On the other hand, he deals with infinitesimal calculus in the 17th *Lettre philosophique*. In doing so, he could rely on Fontenelle's *Éléments de la géométrie de l'infini* or the writing *Commercium epistolicum*, in which Newton claims against Leibniz to be the first inventor.[54]

The French Encyclopedists also emphasize the importance of mathematics. Diderot praises the usefulness of mathematics for the sciences and for scientific thinking, as it can systematically derive one thing from another and thus find something new.[55] D'Alembert describes in his preface that the aim of the encyclopedia is to present the arts and sciences in their current state and to show the historical steps that have led to this. In his preface, he places particular emphasis on the era from the Renaissance to the end of the 17th century and the period from 1700 to 1750. He notes that the development from the scholarship of the Renaissance, through the brilliance of the literary works of the 17th century, has led to the current flourishing of philosophy.[56] That he understands philosophy in a very broad sense becomes clear when he gives examples. Thus, he praises the merits of Bacon and Descartes for mathematics and physics and the achievements of Galileo in astronomy, Harvey regarding the circulation of blood, Huyghens in geometry and physics, Pascal on air pressure and combinatorics, Malebranche on sensory illusions, Boyle on experimental physics, and finally Vesalius, Sydenham, and Boerhaave in medicine. With Leibniz he highlights the merits in calculus, but considers his teachings on monads and pre-established harmony less wise than the theories of Locke or Newton.[57]

In this context, the scientific texts of Fontenelle or Buffon make use of literary means. However, fictional literature generally did not reach the level in the 18th century that it had in the previous century. D'Alembert particularly highlights Francis Bacon, as he owes him the division of the fields of knowledge in a tree of knowledge. He also emphasizes the necessity of experimental physics, limits the sciences to the useful, and recommends the study of nature and the comparison of experiments. Descartes appears to him great as a representative of algebra and geometry. Thus, his theory of light refraction is to be understood as an application of geometry to physics. As an inventor, he

---

[53] Stenger 2013, 14.
[54] Stenger 2006, 15.
[55] Diderot 1975, 365, 367.
[56] Alembert 1894, 76.
[57] Alembert 1894, 107.

also stands out in his astronomical considerations, which associate planetary movements with fluids and centrifugal gravity. As a philosopher, d'Alembert believes, Descartes was less successful, although he is at least credited with having shaken the authority of scholasticism and debunked old prejudices. He also sees Newton's merits in explaining the planetary orbits by gravitation and in optics. His physics is experimental and geometric, "uniquement soumise aux expériences et à la géométrie."[58] John Locke is credited with reducing metaphysics to an experimental physics of the soul, a "physique expérimentale de l'âme." Like experimental physics, it should carefully collect and summarize facts, explain one fact through the others, and highlight those facts that serve as a basis for the others.[59]

The experimental natural sciences demonstrate their progress through ever new inventions. If one wants to gather the criteria for evaluating inventions, it is worth considering the beginnings in antiquity. A look at the Greek tradition of inventor directories initially shows the high rank that was attributed to inventions. After all, the gods were initially named as the first inventors. Even if the mythical inventors of ancient times are eventually replaced by inventive people of the recent past, it remains to be noted that even today the inventors of the gray past, who invented the axe, spear, bow and arrow, the oil lamp, and spinning, are unknown. A genetic and mythological perspective is brought by Aeschylus with his play *Prometheus* in the 5th century, in which the title hero brings not only fire, but the entire culture to humans. The historian Herodotus sticks to the mythological explanation. He traces the origin of the Greek characters back to Cadmus, who brought the letters from Phoenicia, so that the Ionians would have received them via Thebes. In the 5th century, the pre-Socratic philosopher Xenophanes distinguished between the divine and the human part, claiming that the gods had not shown all the hidden things to mortals from the beginning, but that humans gradually found the better while searching. With the appearance of the Sophists, the idea spread that not gods, but humans are the inventors of cultural goods, which has special significance for the assessment of laws. If these were not imposed on humans by the gods, but are human inventions, then there can be discrepancies between different human laws.

Daedalus is the prototype of the technical inventor. However, the Greek catalogs also listed objects as inventions that go far beyond technology and affect general social phenomena such as politics, trade, and olive cultivation. In this case, not historically tangible inventors are named, but gods, heroes, and peoples. The indication of inventors means the insight into the historicity of these inventions. They create a tradition with the invention. Thus, Pliny offers in his catalogs arranged by disciplines the history of the alphabet, shipbuilding, and music. However, he resigns in the case of writing. Since its

---

[58] Alembert 1894, 100.
[59] Alembert 1894, 104.

origin dates back to ancient times, a historical origin can no longer be given and an *aeternus litterarum usus* (eternal use of letters) must be assumed. It is not always the first ones who are mentioned. It can also be those who have later emerged as outstanding representatives, so that not the absolute beginning, but the particularly visible and representative in the present is mentioned. This was particularly important for Greek cultural apologetics in order to relativize the dependence of Greek culture on Egyptian culture. After all, the Egyptians were often seen as the first inventors, e.g., of astrology and geometry. Here, the Greeks had attributed to themselves the advantage of a special *Metis,* an inventiveness peculiar to them.

Plato has Protagoras tell in the dialogue of the same name about a time when there were already gods, but no mortal beings yet, who were later formed by the gods from earth and clay. Epimetheus equipped them. When he showed Prometheus his work, all the advantages and abilities had already been distributed to the animals, so that man remained without means. Prometheus did not want to tolerate this, and saved him by stealing the arts of Hephaestus and Athena and giving them to him. Now he had conveniences through agriculture, dwellings, clothes, language, and statecraft. The need (*chreia*) now becomes the principle of culture creation. The three most important needs are food, housing, and clothing. The satisfaction of needs is the reason for the creation of areas of knowledge and culture. The increasing refinement of needs requires an increasing differentiation of knowledge areas. A comparable purpose orientation is later also shown in Aristotle in the doctrine of the four causes of things.

The invention refers to something that was not there before, so to something new. However, when one retrospectively deals with inventions in history, one is discussing the origins of objects and areas from technology, science, and culture. The history of inventions is a history of beginnings. Where the invention exceeds the field of technology, it leads to theories of cultural origin. Where it exceeds the field of artifacts, human products, it leads to general theories of origin. In the representations of inventors and inventions, historical circumstances, utility, purpose, and benefits that the inventions bring are also cited. This outlines the discursive environment as a condition of the inventions. Evaluating and contextualizing in texts about inventions falls into the realm of discourse history.[60] The demonstration of the changes that have resulted from the invention, for better or worse, testifies to an optimism or pessimism about progress. Since inventions initiate traditions, dealing with them is also a kind of genealogy[61] of traditions. Thus, the consideration of the invention of medicine becomes a reflection on its original meaning and function. Therefore, the analysis of the representations of inventions and innovations can also be described as historical heuristics.

[60] Foucault 1969, 249–250.
[61] Foucault 2002, 166–167.

In antiquity, the postulate of utility associated with invention was generalized to the idea of a universal purposefulness. In the *Memorabilia* of Socrates' student Xenophon (430–354 BC), the argument is that works that bring benefit are not results of chance, but of rational considerations. Thus, in humans, eyes, ears, nose, tongue, and hands have specific functions. With his hands and his mind, man surpasses the animals. The world is arranged for him, everything is best ordered by the gods for his sake. The principle of all things is the utility for man.

Entire disciplines of knowledge and tasks and activities assigned to social classes are justified by the narrative of their invention. What legitimization strategies are used, for example, in the writings on military art? Luis Pacheco (1570–1640) goes into great detail in his book on fencing. He quotes Aristotle, according to whom all things that exist want to exist and therefore provide for the preservation of their own existence. From a syllogism, it can be concluded for the human species that all wish for a long life and need to be protected from attacks by others:

> "como el hombre, por ser mas noble que todos ellos juntos, tuviesse mas necessidad de conservarse: y muchas vezes (que es harto dolor y lastima) fuesse ofendido de sus semejantes: pues, como dize el adagio: El hombre es lobo del hombre, fuele necessario un arte que le enseñasse como auia de hazer esta defensa, que le sirviesse de amparo, contra un enemigo tan poderoso, de tantas fuerzas, y de tanta malicia como el propio hombre."[62]

And to improve the defense against deceit and envy, the art of war was invented. As a science, it serves to defend the peaceful, peace-loving people and should protect them from suffering. In his chapter "De como en el mundo fue hallada la Milicia", Scarion imagines an original state in which troublemakers caused unrest until the military was invented and virtuous as well as defensive soldiers with weapons defended the weak who were exposed to attacks and insults. They took on "el cuidado de defender los aldeanos y villanos de los agrauios y injurias, que les hazian los malos y ruines hombres."[63] The problem of threat became a task, for the solution of which the military was invented. Here too, necessity and need form the starting point: "la milicia era necessaria para destruir la malicia humana, y alcancar el bien de la paz y la quietud del vivir humano."[64]

Are contemporary inventions viewed differently than those of the past? Do they prove the superiority of the present over the past? The above-mentioned Francis Bacon particularly highlights three inventions of recent times: the printing press, gunpowder, and the compass. These three have changed the shape of things and human conditions on earth; one in the sciences, the

---

[62] Pacheco 1605, 62.

[63] Scarion de Pauia 1598, 63.

[64] Scarion de Pauia 1598, 64.

other in warfare, and the third in navigation. Countless changes have followed them, and no dominion, no sect, no star seems to have ever exerted greater influence on human relations than these mechanical things, which also prove the superiority of the present over antiquity. When it comes to experience, Bacon distinguishes the new experience from the old, which was merely vague and oriented towards what perception happens to encounter. Bacon contrasts this with the experience controlled, directed, and targeted by the mind, namely the experimental experience.[65] Jean Bodin considers in a similar way in his 1566 published *Methodus ad facilem cognitionem historiae* the inventions of antiquity only as preliminary achievements. According to him, Charles Perrault also highlights in his *Parallèle des anciens et des modernes en ce qui regarde les arts et les sciences* (1688–1696) the merits of the first inventors, who occupy a rank between humans and gods:

> "J'avoue que c'est une grande louange et un grand merite aux Anciens d'avoir esté les Inventeurs des Arts, et qu'en cette qualité ils ne peuvent estre regardez avec trop de respect. Les Inventeurs, [...] sont d'une nature moyenne entre les Dieux et les hommes, et souvent même ont esté mis au nombre des Dieux pour avoir inventé des choses extremement utiles."[66]

But if one compares the old inventions with the new ones, Perrault considers the latter more powerful, as in the case of knitting, where the machine is superior to the most skilled hand, as it accomplishes in a moment the various movements that the hands make in a quarter of an hour.[67] Thus, Perrault, like Bacon, also pays special attention to new inventions such as the compass, printing press, and gunpowder.[68]

It thus becomes apparent that progress in France is associated with the new, scientifically oriented English models of thought. Voltaire praises Bacon's experimentation and Locke's logic. D'Alembert describes the development from the Renaissance to his present as progress and emphasizes the necessity of experimental physics, the comparison of experiments, and the limitation of the sciences to the useful, citing Bacon.

Since antiquity, it has been inventions that made progress recognizable. The invention of new techniques and fields of knowledge stimulated reflection on their original meaning and function. Thus, disciplines of knowledge and associated activities, such as warfare, are legitimized by the narrative of their invention. When comparing old inventions with new ones, the latter appear more significant. This is shown by the contemporary new inventions such as the compass, printing press, and gunpowder for both Perrault and Bacon, thereby demonstrating the progress achieved.

---

[65] Schmitt 2002, 17.

[66] Perrault 1964, 119.

[67] Perrault 1964, 120.

[68] Perrault 1964, 395, 401, 443; Strosetzki 2022, 20–34.

5.5   DECLINE

Most French Enlightenment thinkers oriented themselves towards progress, which they admired in the English models of scientific and empirical thinking. However, there is also the counter-position, which asserts that sciences, particularly natural sciences and technology, have not contributed to progress, but have brought about a loss of good ways of life. The most prominent representative of this thesis is Rousseau. In his second discourse, he provides an answer to the question of the Academy of Dijon, what the origin of inequality among men is and whether it is compatible with natural law. He proceeds in an apparently historical manner, imagining the *homme originel* or the *homme naturel* as he must have emerged from the hands of nature. Since man is good in his origins but bad in the present, the question arises as to what has corrupted him. For Rousseau, it is the optimization of human understanding and the process of socialization. He does not see the fact that the mind originally distinguishes man as an *animal rationale* from all other living beings and is precisely his advantage as a contradiction. For it is the conveniences that harm him. When he had no house and no clothes, he was more resistant. Without an axe, his fist was still strong. Without a sling, he could throw further and without a horse, he could run faster. While pity used to moderate the instinct for self-preservation, the mind later turns it into an unnatural selfishness that limits his interest to himself and makes others indifferent. Only when man encloses a piece of land and says it belongs to him, when property is created, does a series of crimes, wars, and sufferings begin. He eventually becomes a fisherman, hunter, and warrior. Technological advances lead to weapons and tools. His superiority over nature makes him proud. A division of labor develops between man and woman. Nations emerge. However, with the invention of weapons and tools, his physical strength decreases and he loses his previous lack of needs. As a result of these external changes, man must leave his original state. Evils and vices arise. Blacksmithing and agriculture bring further developments. The blacksmith, as the first man who needs food but does not produce it, forces another to produce for two. Division of labor brings dependence and the loss of natural equality. Agriculture, in turn, requires the division of the land, which previously belonged to everyone, as one wants to be sure to be able to harvest in the fall what one has sown in the spring. With property, laws are created. The state's positive law replaces natural law. The laws secure property and perpetuate inequality, as they assure the rich of further power. While the savage lived freely in his inner self, the civilized man is oriented towards the opinions of others.[69]

Rousseau's originality has often been questioned. Diderot claimed that Rousseau continued Seneca's 90th letter to Lucilius in his second *Discours*. And finally, the ignorance towards the advances in sciences had been defended a hundred times before him. Not least, the praise of the folly of

---

[69] Bosshard 1967, 20, 45, 47, 5, 8.

Erasmus of Rotterdam could be mentioned here. Montaigne had already regretted that artificial inventions threaten to suffocate nature. Seneca discusses in his 90th letter the development of mankind from its beginnings to the present. The first humans followed nature uncorrupted. To his misfortune, man has distanced himself from nature in the course of his development. Once happy and carefree, he is now restless and anxious. Once he had everything he needed, but now he is poor because he never has enough. Once free, socially perfectly integrated into the community, he is now a slave. The physical lack of needs resulted from the natural measure of the necessary, as there was neither agriculture nor metal, nor precious stones. People slept well on the hard ground, the dense forest protected them from sun, storm, and rain. The decay of the original state began with greed, which set in with property, according to Seneca. For when personal property was created, people stopped owning everything. In addition, there were desires for unnecessary and unnatural things, so that the mind began to serve the desires of the body.

It thus becomes apparent that lack of needs characterizes the natural man in Rousseau as well as in Seneca. Technology, which softens through conveniences, and property lead both away from the natural state. Both reject technology. However, Rousseau's natural solitude is contrasted by Seneca's natural community with a ruler, which leads to a natural inequality in Seneca, for whom the foundation of states is not a bad tool of oppression, as the state has always existed.

Voltaire wrote a satirical dialogue titled *Timon* in response to the publication of Rousseau's *Discours sur les Sciences et les Arts* in 1751. Timon tells an acquaintance that he has burned all his books because they corrupt people. Especially the books on mathematics and geometry are a scourge of humanity, as without them we would still be living in the golden age. Arts and sciences are responsible for the constant emergence of new needs and passions that lead to crime. If Cardinal Richelieu had not been a theologian, he would have fewer murders on his conscience. And if the Catholics in Ireland had not been so intensively engaged with the works of Thomas Aquinas, more Protestant families would still be alive. When the two dialogue partners are robbed by thieves on their way to an invitation, Timon asks them at which university they had studied. They reply that they cannot even read. And when they arrive at their hosts', who are among the most educated people in Europe, penniless and barely clothed, Timon, following his premises, had to fear being strangled. Although the opposite happened, after the meal Timon asks for pen and ink to polemicize against education, "pour écrire contre ceux qui cultivent leur esprit."[70]

When Voltaire was sent Rousseau's second *Discours*, he thanked him in a letter dated August 30, 1755. He felt the urge to walk on all fours again and

---

[70] Voltaire 1785, 16.

live with the indigenous people in Canada. However, this was impossible for him because he depended on modern European medicine and there was war there. An indigenous Canadian is also the protagonist of his novel *L'Ingénu* (1767), who comes to France full of naivety and encounters civilization there. In prison, he meets a teacher who instructs him in physics, natural sciences, and philosophy. The myth of the noble savage, who, however, develops into a fearless warrior and philosopher, is in the background. Here, it is the natural sciences that are considered particularly crucial for the progress of civilization. Therefore, criticism of civilization primarily targets them.

It is advancing civilization that, according to Rousseau, leads to softening and the loss of the strict morals of the ancestors. The Battle of Thermopylae, where the Spartans faced a numerically far superior Persian army, fought heroically but were defeated, is often cited as an ancient example of heroic morals. Since then, military skills have been attributed to the Spartan's simple lifestyle and distance from luxury. According to Xenophon, Spartan education includes boys not wearing shoes and being toughened by wearing the same cloak in summer and winter. The Spartan king Agesilaos was also a model of lack of needs. The criticism of luxury among the Spartans is not least in contrast to the so-called barbaric peoples, especially those of the Orient, and particularly Lydia, located on the Mediterranean coast of Asia Minor, with its proverbially rich King Croesus.

It is primarily technical advances that make life more comfortable than it would originally be without them. But since they remove us from nature, they can also be considered unnatural. If one raises vital or natural needs as a standard, then deviating from it is against nature, as Seneca believes: "Omnis vitia contra naturam pugnant, omnia debitum ordinem deserunt; hoc est luxuriae propositum."[71] Seneca distinguishes between natural desires, which have limits, and unnatural ones, which know no bounds. While natural desire can stop somewhere, unnatural desire wanders around without limits. The good life is according to Aristotle not achieved through luxury, but through virtuous living, for even with moderate means one can act according to virtue. This can be clearly seen in the fact that private individuals do not seem to be inferior to princes in right and virtuous action, but rather seem to be ahead. So it is enough if the necessary means are available.[72] The cardinal virtue of *moderatio* advises according to Aristotle to choose the middle measure, that middle, namely, which, according to our conviction, corresponding to right reason, lies between excess and deficiency.[73]

However, one can also like Voltaire negatively evaluate the original natural conditions. His verse satire *Le mondain* and its sequel *Defense du Mondain ou l'Apologie du Luxe* (1736–1737), which he wrote after a

---

[71] Cf. Grugel-Pannier 1996, 27.

[72] Aristotle 1995b, 254.

[73] Aristotle 1995b, 130.

two-and-a-half-year stay in England, are a critique of an original state. When Voltaire imagines Adam and Eve, the image is not very flattering, as they had to sleep on the floor and lacked the tools to cut and clean their fingernails: "Avouez-moi que vous aviez tous deux / Les ongles longs, un peu noirs et crasseux, / La chevelure un peu mal ordonnée, / Le teint bruni, la peau bise et tannée. […] Le repas fait, ils dorment sur la dure: / Voilà l'état de la pure nature."[74] Therefore, Voltaire praises in the article *luxe* of his *Dictionnaire philosophique* the invention of scissors for hair and fingernails as well as that of the shirt. In summary, even though luxury has been condemned for 2,000 years, it has always been popular: "On a déclamé contre le luxe depuis deux mille ans, en vers & en prose, & on l'a toujours aimé."[75]

Looking back on such debates, the Scottish empiricist David Hume (1711–1776) distinguishes two attitudes towards luxury. While strictly moral thinkers condemn even the simplest luxury as a cause of corruption, liberally minded spirits even praise the vicious luxury and highlight its benefits for society. Even if luxury may be immoral, it seems good to the liberals at least for the economy. For the author of the *Dictionnaire historique et critique* (1697), Pierre Bayle (1647–1706), it has the advantage of contributing to the circulation of money. From a societal perspective, luxury serves the poor more than the rich, as the French author Gabriel-François Coyer (1707–1782) emphasizes. Even if luxury ruins the rich houses, it still feeds producers and manufacturers. And for the political theorist Montesquieu (1689–1755), luxury is even a contribution to poverty reduction: "Il faut bien qu'il y ait du luxe. Si les riches n'y dépensent pas beaucoup, les pauvres mourront de faim."[76] In the balance of moral and societal value of luxury, the latter seems to him more weighty.

The social theorist Bernard Mandeville (1670–1733), who grew up in the Netherlands and lived in England, separates morality and economy. His addressee is not the prince, but the individual. He thus shares Bayle's and Montesquieu's opinion that what is morally questionable can be socially beneficial. For Mandeville, luxury is everything that goes beyond the necessary livelihood of man, which leads him to reject the basic state as primitive simplicity. He sees the latter given in an a priori conceived natural state, while the societal state already has luxury goods. He therefore rejects frugality as the principle of keeping away from all superfluities, as it leads to primitiveness. The absence of luxury in Sparta is only the flip side of the oppressive military service. Spartans are characterized only by the lack of needs; the amenities of civilized countries are unknown to them as well as the arts. What happens to a wealthy country from which greed, avarice, and luxury are banished, Mandeville illustrates with his 1705 published and variously expanded Fable

---

[74] Voltaire 2003, 298–299.

[75] Voltaire 1764, 256; see also 258.

[76] Strosetzki 2022, 175–186, 182–183.

of the Bees. While the state flourished through the vices of individuals, the situation changes after Jupiter makes pride, luxury, and crime disappear: The societal productivity languishes, numerous professions become superfluous and unemployed bees leave the state.[77] It was the vices of the individual that maintained societal prosperity.

It thus becomes apparent that science and technology can indeed bring relief. But where they exaggerate and lead to luxury, they move away from the natural and lead to the loss of good ways of life, which can be assessed as decadence. Critics of this primarily moral decline are opposed by those who, for their part, do not judge the development as a decline, but point out the disadvantages of a state of nature.

---

[77] Grugel-Pannier 1996, 198–200, 214, 216, 242–243.

PART III

PRIORITY OF THE NATURAL SCIENCES

CHAPTER 6

Procedures

6.1 INDUCTION

Voltaire praised progress using the English model as an example. Indeed, scientific methods were introduced and established by English philosophers. Induction, which starts from individual observations, generalizes them, and derives general knowledge from them, replaces deduction, which takes the reverse path from the general to the specific. The rejection of deductive and the preference for inductive approaches made quantitative surveys based on given materials relevant. The thinking and acting subject does not derive its knowledge from introspection, but from the observation of objects. Teleology is replaced by causality.

Francis Bacon (1561–1626) is considered the founder of an empiricism shaped by observation and induction. The title of his work *Novum Organon* (1620) announces the attempt to overcome the *Organon* of Aristotle, which deals with categories and logical conclusions. Starting from observations and experiments, one should methodically proceed first to sentences of lesser, then to those of higher generality. For him, not formative forms *(formae substantiales)* or purposes explain developments as in Aristotle, but laws that are to be sought in nature. Therefore, he rejects the idea that man is a microcosm in analogy to the macrocosm of nature and carries the fundamental ideas within him. Physics must free itself from anthropomorphism, which assumes goal and purpose causes in nature as in humans, instead of deriving the causes of phenomena from material necessity. He criticizes Plato and his school, as they start from abstract forms, final purposes, and first causes.[1] Knowledge serves to expand human power through new inventions. Thus, knowledge is

---

[1] Bacon 1999, 135.

© The Author(s), under exclusive license to Springer-Verlag GmbH, DE, part   149
of Springer Nature 2025
C. Strosetzki, *Literature in Dialogue with the Natural Sciences*,
https://doi.org/10.1007/978-3-662-71319-8_6

not an end in itself, but serves a purpose. One can only do as much as one knows.

Bacon notes that both in Greek and Roman antiquity and in the later Christian era, natural philosophy was neglected, even though it is the great mother of all sciences. Other disciplines are dressed up and made up, but without growth power. Thus, he laments that with the beginning of the Christian era, the brightest minds were occupied with theology, as the most rewards were offered in this area and all kinds of aids were provided. In antiquity, moral philosophy played the role of religion among the pagan Romans, but the elite of that time turned to political tasks, as the size of the Roman Empire required the participation of many people. Among the ancient Greeks, interest in nature was only short-lived. Bacon saw the astronomical considerations of Thales soon replaced by Socrates and Plato, who in turn turned to moral philosophy and civic affairs. Scientific thinking thus appears to Bacon to have been hindered so far by the preference for non-scientific disciplines: Therefore, if natural philosophy was greatly neglected or hindered in those three periods, it is no wonder that people made little progress in it, as they were doing something completely different.[2] Bacon sees another obstacle to natural research in misunderstood religiosity. Just as the Greeks attacked those who saw natural causes in lightning and storms, not the work of the gods, it was Christian clerics who fought those who believed, based on supposedly secure evidence, that the earth was a sphere. The latter Bacon attributes this to the fact that in the Middle Ages theology was mixed with the philosophy of Aristotle. This had the consequence that the new, even if it was better, was not only rejected, but immediately eradicated. There was fear that in the exploration of nature something might be discovered that would overthrow or shake religion. Against this background, it does not surprise Bacon either if the growth of natural philosophy has been inhibited.[3]

Bacon rejects pure experience just as he does pure rationality. He compares empiricists to ants, who only collect and consume, and rationalistic dogmatists to spiders, who create their webs from within themselves. In contrast, he recommends the method of bees, who draw the juice from the flowers of gardens and forests, but then process and digest it with their own strength. Thus, natural science should also transform and process what is gained from experiments in the mind.[4] Bacon does not want to derive experiments from experiments like the empiricists, but from the works and experiments the causes and principles, and from these two again new works and experiments.[5] He wants his inductive method to be applied not only to natural science, but also to other sciences such as ethics and politics. It is comparable to its

---

[2] Bacon 1999, 171.

[3] Bacon 1999, 200–201.

[4] Bacon 1999, 211.

[5] Bacon 1999, 243.

counterpart, the syllogism of logic, which does not limit itself to the natural sciences, but extends to all sciences.[6] Only when the mind is linked with things is it true that the art of inventing can strengthen with the inventions.[7] In summary, Bacon maintains that man can make nature useful through various work and not through disputes or useless magical formulas, from which the desirable result, an improvement of human conditions and an extension of his power over nature,[8] arises. He does not want to stop at the observation and knowledge of nature, but sees his goal in active activity and world change.

Following Bacon, John Stuart Mill (1806–1873) relativized the importance of deduction. As is well known, a syllogism has a general major premise from which the conclusion is derived. From the major premise "All penguins can swim" and the minor premise "Abel is a penguin," the conclusion can be drawn: "Abel can swim." But if one asks how one arrived at the major premise "All penguins can swim," it becomes apparent that this can only be obtained from numerous singular sentences like "Penguin Abel can swim," "Penguin Bernhard can swim," "Penguin Caesar can swim." Against this background, according to Mill, deduction proves to be a secondary procedure based on the truth of inductive sentences. Deduction seems to only summarize the singular sentences from which its major premise was obtained.[9]

In summary, it can be stated that the English empiricists, so much admired by Voltaire, proceed inductively based on experiences and experiments, i.e., they generalize concrete individual observations and thus move from sentences with a lower degree of generality to sentences with a higher degree of generality. Similarly, concepts are not derived from substances, but are obtained through generalization. While in antiquity all attention was paid to ethics and politics and in the Middle Ages people were occupied with theology, now the exploration of nature with the inductive method should be at the center, which in turn should determine ethics and politics. For Bacon, knowledge as such is not sufficient; knowledge serves to improve human conditions and to extend power over nature. When John Stuart Mill criticizes the ancient deductive method, since it has obtained its major premises through induction, he is certainly right in terms of the approach to knowledge. However, what remains open with him is the question of whether the general major premise might not have a validity and truth that is independent of the method by which it was obtained.

---

[6] Bacon 1999, 263, 265.

[7] Bacon 1999, 275.

[8] Bacon 2009, 611.

[9] Lüthe 1998, 163–164.

6.2   EXPERIENCE AND EXPERIMENT

The pioneer of the Enlightenment John Locke (1632–1704) was the one who expanded Bacon's empirical approach, as he also denied the existence of innate ideas and propositions. For him, the mind is originally without content, until it arrives at knowledge through perception. Experience provides him with the material for his reason and his knowledge. Through observation, which is directed at external perceptible objects, he grasps these external objects, and through the intuition of the inner senses psychological processes. In this process, extension and spatial determinations belong to the objects themselves, while colors and tones are sensory qualities of the perceiving subject. Only from this point are simple and complex ideas formed. Genera are nothing more than summaries of many similar things by the respective word. In his *Essay Concerning Human Understanding*, which was drafted in 1670 and fully published in 1689, Locke sets out to investigate the origin, certainty, and extent of human knowledge, along with the foundations and gradations of belief, opinion, and consent.[10] The mind is comparable to the eye, which sees and perceives all other things without being aware of itself. Therefore, it is all the more important to make it its own object with a certain distance.

If one follows the development of a child from birth, one sees how the mind develops. As the mind is supplied with ideas through the senses, it has more material to think about. After some time, it begins to recognize objects with which it has to deal repeatedly. It learns to retain and distinguish the ideas supplied to it by the senses. Gradually, it perfects itself in this area, until it then expands its notions, combines them into ideas, and abstracts them.[11] However, caution is advised with incorrect idea connections. The false connection of actually unrelated and independent ideas in the mind has the power to direct actions and passions in the wrong direction, which is why parents and educators should warn against it. An example is that ghosts and goblins are always associated with darkness, with which they have no more to do than with light. Therefore, a child will be afraid of ghosts in the dark. Another example is an insult that one person inflicts on another. This results in the future association of insult and insulter, and the person who once insulted will be met with hatred and aversion in the future.[12] So, in Locke's view, epistemological empiricism has consequences for human behavior.

These become even more apparent in the works of the Scottish Enlightenment thinker David Hume (1711–1776). For him too, concepts that make sense should be traceable back to the sensory impressions from which they derive. Particularly important to him is the uniformity hypothesis, which he associates with habit, which is nothing more than an expectation

---

[10] Locke 2006, 22.

[11] Locke 2006, 124.

[12] Locke 2006, 502–503.

for the future formed on regular experience. Thus, it is to be expected that the sun will rise again tomorrow because it has always risen in the past. There are no innate ideas of substance or essence. Ideas only have meaning if they can be traced back to empirical impressions. Conclusions are based on the relationship between cause and effect, which can be illustrated by the example of billiard balls. If one is lying on the table and another is moving towards it at a certain speed, then they collide, with the ball that was stationary now being set in motion. For causes to have effects, therefore, a touch in space and time is required. In addition, the cause precedes the effect in time and the same process always happens when the experiment is repeated, i.e., the correlation is constant. According to Hume, all philosophy is based on such conclusions, which we know from practical life.[13] All causal conclusions are therefore based on experience and on the assumption that the course of nature will remain uniformly the same without interruption. If there are no substances, then the concept of a body, such as a peach, is nothing more than the composition of the ideas of a specific taste, a color, a shape, size, and consistency. The same applies to the soul, which is composed of perceptions such as heat and cold, love and anger, or thoughts and sensations, without there being anything continuously identical or a substance.[14] If he denies on this basis that there is a once and for all established difference between the morally right and the wrong, he exposes himself to the accusation of destroying the foundations of morality. He tries to refute this by bringing into play not only a natural inclination but also the general interests of human society and the context of action with other people.[15]

The founder of classical national economics, Adam Smith (1723–1790), is the one who further contemplates these consequences for ethics with his *Theory of Moral Sentiments* (1759), also starting from experiences and sensations. He strives to solve the problem of how to inductively and through observation arrive at ethical norms that are not deductively derived from universally valid principles and yet are not relativistic. When he asks what virtue consists of or what behavior deserves moral approval, he is seeking the criterion of morality. When he asks about the principle that makes us perceive certain actions as morally valuable and others as morally worthless or reprehensible, he is concerned with the foundation of morality. He answers both questions with the construction of an impartial and well-informed spectator imagined by us. This impartial spectator alone teaches us the real insignificance of our own self and everything that concerns us, and only through the eye of this impartial spectator can the natural deceptions of self-love be corrected. He shows us the beauty of generosity and the ugliness of injustice; he shows us how beautiful it is to renounce the greatest personal advantage and

---

[13] Hume 1980, 23.

[14] Hume 1980, 47.

[15] Hume 1980, 121.

sacrifice it for the even greater interest of other people.[16] On the other hand, the traditional virtue of prudence is redefined as care for the health, wealth, rank, and reputation of the individual, i.e., for the things that, according to general opinion, his well-being and his felicity in this life primarily depend on.[17] The criterion is thus empirical and purely formal. A morally good action is one that still appears good to us when we view it from the stand-point of an imagined spectator. The decision is derived not from reason, but from feeling. Sympathy is the ability to put oneself in the role of others and judge whether their motives are decent or selfish. Smith gives as an example a joke, where we approve of others' laughter, even though we ourselves do not laugh, perhaps because we are in a bad mood. Experience has taught us, however, what kind of jokes are capable of making us laugh in most cases, and we notice that this is such a case. Therefore, we approve of the laughter of society and feel that it is quite natural and appropriate to the occasion.[18] Since sympathy here does not mean benevolence, the model is quite compatible with self-love and egoism and does not contradict Adam Smith's economic theory. He certainly abandons an innate idea of the good, as Plato and Aristotle postulated.

If Bacon already wanted the inductive method to be applied to ethics and politics, then Adam Smith shows with his imagined impartial and well-informed spectator how to imagine this more precisely, with morality becoming the result of a special kind of empirical observation. If knowledge as such was not sufficient for Bacon, since knowledge serves to improve human conditions and expand power over nature, then in view of the uniformity hypothesis with David Hume, it allows future predictions. Repetitions of experiences are for Hume what enable laws and concepts, with the latter not being substances, but like the peach or the soul, compositions of ideas. John Locke had illustrated this using the development of the child, who gradually assembles and abstracts his sensory notions into ideas. In doing so, false ideas, such as that ghosts are always associated with darkness, should be corrected.

6.3   Language Criticism

If natural languages are suspected of leading to pseudo-problems due to their ambiguities and inaccuracies, then they are subject to a critical analysis that has nothing to do with content, but with form. Language criticism often targets speculative philosophical directions. However, insofar as it rejects any non-exact language, it also targets literature. An early form of language criticism was practiced by the skeptics when they doubted not only knowledge but also the communicability of knowledge. There were further precursors

---

[16] Smith 2010, 215.

[17] Smith 2010, 344–345.

[18] Smith 2010, 21.

of language criticism in the Middle Ages. When the nominalists saw in the general concepts no longer essences, but only names, they turned against Platonic conceptual realism. The English empiricists followed suit, warning of the deceptions and misleading aspects of language. According to Francis Bacon, it is often thought that reason uses language, while in reality it is language that dominates the mind so much that thinking is led to useless sophistry. In addition, most words are formed according to the views of the masses. But as soon as a sharp mind wants to change these determinations according to his more precise observation, the words resist, which is why disputes among scholars often degenerate into quarrels about words and names. It would be better, Bacon thinks, to start with definitions like the mathematicians. Bacon distinguishes between names without things and names of things that really exist, but are confused and inappropriate. He counts luck, the first mover, the spheres of the planets, the element of fire, and similar inventions that have arisen from vain and false doctrines among the first group. This kind of idols can be more easily eliminated, as they can be eradicated by persistent denial and by setting aside the doctrines.[19] For the second group, he uses the word "moist" as an example, which easily escapes in all directions, easily divides and disperses, easily binds and gathers, easily flows and consists in movement, easily sticks to another body and makes it wet, easily becomes liquid again or also flows together if it was previously solid.[20] The word thus proves to be a confused sign for various effects and is to be viewed as critically as the words of the first group.

Locke practices even more extensive language criticism. He expresses his astonishment that indefinite and meaningless phrases were considered secrets of science and difficult-to-understand, misused words with little or no sense were valued as deep scholarship through long habit.[21] He takes up Bacon's example of the word "moist" and tells of a doctors' meeting where the question arose whether a fluid penetrates the fibers of the nerves. After a long controversial discussion of the pros and cons, the suggestion was taken up to first examine what the word "moist" actually means. They found that the word was not as precisely defined as everyone had assumed, and everyone had understood it differently. They realized that the meaning of that expression had been the crux of the dispute and now saw that their opinions about the fluid matter in the nerve channels themselves differed only slightly.[22]

Different peoples have different customs, habits, and manners, which are named with words, which explains why some words exist in one language that are missing in another. The change of habits and opinions brings forth new

---

[19] Bacon 1999, 123.

[20] Bacon 1999, 123.

[21] Locke 2006, 11.

[22] Locke 2018, 112–113.

words.[23] Thus, one can find numerous words in one language that do not exist in another language.[24] Depending on the thinking habits one is familiar with, one believes that certain words correspond exactly to the nature of things. Thus, the Platonist considers the world soul to be really existent and the Peripatetic the ten categories or substantial forms. If air and space vehicles played a central role in any doctrine, the follower of this doctrine would naturally believe that they really existed.[25] Where would there be a scientific dispute or a casual conversation about honor, faith, grace, religion, church, etc., where the difference in understanding that people have of these concepts would not be easily observed?[26] This means that they have different opinions regarding the meanings of these words.

In communications through words, Locke distinguishes between a civil and a philosophical use. The former serves ordinary oral interaction and exchange of opinions about everyday matters. The latter claims to express secure and undoubtable truths in general sentences, on which the mind can rely.[27] Figurative expressions and wordplays bring pleasure and amusement and are ornaments of speech, therefore they cannot be described as imperfection or misuse of language. However, they serve no other purpose than to introduce false ideas unnoticed, to arouse passion, and thereby mislead judgment.[28]

In Germany, it was Johann Gottfried Herder (1744–1803) who, in the spirit of English empiricism, demonstrated how pseudo-problems arise from a misuse of language and rejects these as transcendental steam.[29]

When the members of the Vienna Circle published the journal *Erkenntnis* from 1930 to 1938, they wanted to overcome pseudo-knowledge by giving logic in philosophy the place that mathematics had in physics. The group included Rudolf Carnap, who criticized the lack of progress in knowledge in metaphysics compared to the natural sciences. He suggested that philosophy should no longer make statements about the world, but analyze the statements of the individual sciences and deal with the difference between meaningful and meaningless sentences. After all, the real world is the subject area of the individual sciences, for which every sentence is meaningful that is either true or false and can be verified. Rudolf Carnap distinguishes in *Logische Syntax der Sprache* (1934) between an object language, which refers to real objects, and a meta-language, whose object is the object language. The place of philosophy should be the meta-language, which refers to the

---

[23] Locke 2006, 360.

[24] Locke 2018, 41.

[25] Locke 2018, 129.

[26] Locke 2018, 106.

[27] Locke 2018, 101.

[28] Locke 2018, 144.

[29] Herder 1969, Vol. 2, 282.

object language and not to the objects themselves. In this context, it should deal with syntax and semantics and work on the construction of languages that can make meaningful statements. However, this view entails a narrowing of the concept of philosophy and thus the renunciation of numerous traditional questions, e.g., of ethics or anthropology.[30]

In his work *Scheinprobleme in der Philosophie* (1928), Carnap (1891–1970) distinguishes three types of sentences. While the first type say something about logical relations and connections, those of the second group are empirically verifiable, and the sentences of the third group are characterized by claiming to say something about reality, but are meaningless because they cannot be verified. Meaningless statements can be recognized by different characteristics. If terms like "world soul" or the invented meaningless word "babig" appear in them, a sentence becomes meaningless.[31] But even with meaningful words, a meaningless statement can be formed if they are syntactically incorrect, like the sentence "Nothing nothings." A statement is also meaningless if it cannot be verified by observation in an intersubjective manner. There are also sentences that say nothing about objective reality, but are expressions of subjective experience. They do not offer knowledge, but refer to emotional states. It is a typical characteristic of literature to write poetry in terms and to convey experiences.

The Vienna-born Ludwig Wittgenstein (1889–1951) suggested in his *Tractatus Logico-Philosophicus* (1922) that one should remain silent about that which is meaningless to speak about. He was referring to philosophical metaphysics. A logical grammar that avoids the ambiguity of everyday language could provide a solution. Self-critically, Wittgenstein does not exclude his own book, as he considers it superfluous and dispensable once it has been understood.[32] In his *Philosophische Untersuchungen* (1945), he no longer views language from the perspective of meaning, but from its use, and calls everything related to linguistic activities "Sprachspiel" (a language game).[33]

Language criticism thus refers to natural languages as they are spoken in everyday life or used by literature and metaphysics. Bacon laments that the words of this language are formed by the masses and are not, as in mathematical language, results of definitions. This is the reason why there are words whose designated objects do not exist and other words, like the word "moist", are ambiguous. Words are created by habits of thought, which is why they are conditioned by these and without these become contentious or even objectless. According to Locke, particularly misleading are metaphorical expressions, which are entertaining but distract from the truth. The Vienna Circle is concerned with the difference between meaningful and meaningless

[30] Guderian 2009, 216.
[31] Carnap 1966, 49.
[32] Wittgenstein 1960, 83.
[33] Wittgenstein 1969, 293.

sentences. Other sentences say just as little about objective reality, as they refer to subjective emotional states.

Wittgenstein and Carnap see the unresolved problems of humanity as results of incorrect language use, which is why they should be eliminated from serious thinking. Philosophy should restrict itself to metalinguistic language criticism and language construction and leave knowledge about reality to the natural sciences. If, therefore, there are no meaningful sentences in metaphysics as in literature, different consequences can be drawn from this by literature. Either one relativizes or refutes these theses or one makes them the starting point of a literary metaphysics.

6.4   Positive Facts

Auguste Comte was born in Montpellier in 1798. He was a tutor for analysis and mechanics, for a time secretary of the utopian social reformer Saint-Simon, and gave lectures in private circles, which he published under the title *Cours de philosophie positive* from 1830–1842. It is Comte's scientific education that shapes his thinking. Since religion and metaphysics as unifying and ordering factors have fallen away, a new authority is needed, which he sees in modern science. Comte refers the word "positive" to the factual as opposed to the imagined. It also refers to the useful as opposed to the idle, if it is aimed at the constant improvement of all individual and collective living conditions, instead of satisfying an unproductive curiosity. Thirdly, it means certainty and thus stands in contrast to the indecision of endless debates that were common in the old ways of thinking. Fourthly, it refers to the precision that corresponds to the nature of the phenomena, and not as before to supernatural authority, which remains uncertain. And finally, the word "positive" can also be understood as a value judgment as opposed to "negative", when it is a matter of not destroying, but organizing. The positive spirit corresponds to general common sense insofar as both their field of application and the experimental starting point and goal are completely identical and both refer to the most everyday phenomena. The basic rule should be that every assertion should be traceable to a fact in order to make sense. Laws can be formed as constant relationships between the observed phenomena. A main characteristic of the positive spirit is foresight, which is a consequence of the discovery of the constant relations between the phenomena and replaces the vain scholarship that only mechanically accumulates facts. The point is to see in order to foresee, and to deduce what will be from what is.

Although Auguste Comte had a polytechnic education as an engineer, he does not see himself as a representative of applied sciences. He does not want to apply the scientific results of others, but to combine ideas.[34] For him, mathematics is a shining example of clarity and precision. If he understands

---

[34] Vatin 2007, 426.

a worldview as positivist, which does not ask for the first causes or last goals like theology or metaphysics, but explains all processes according to the unchangeable laws that underlie them, then mathematics is used as the organon of all other sciences. While the abstract part of mathematics is algebra, geometry and mechanics are concrete sub-disciplines that are based on observation and for which mathematics provides the structures. This also applies to astronomy, physics, chemistry, biology, and sociology, which is understood as social physics, since lawful processes take place in the social body. Psychology should not rely on introspection, but on a physiology that can assign mental properties and states to topologically determined brain areas through observation. His second major work *Système de politique positive ou Traité de sociologie instituant la religion de l'humanité* (1851–1854) aims to build a new politics on the basis of a new religion that makes humanity the highest being.[35]

In the field of ethics, Auguste Comte wants to replace egoism and violence with a new altruism he postulates. For the realm of knowledge, he proposes replacing magical, theological, and metaphysical explanations of the world with positive, scientifically reproducible, and verifiable facts and laws. Nevertheless, he sees the function of religion as a necessary social bond that holds together the otherwise dissolving forces. Religious cult and rites convey to society an image of its unity, which is why they are indispensable for the new positivist society. He refers to humanity with its culture as the "Grand-Être" and now wants this to be revered religiously. Therefore, he compiles a positivist library, a canon of the most important books to read. A positivist calendar, in which each month is assigned a field of knowledge, such as poetry, politics, or modern philosophy, also envisages the veneration of great representatives of humanity, such as Moses, Confucius, Buddha, Shakespeare, Galilei, or Mozart. In this way, it is possible to educate in the spirit of positivism and spread its ideas.[36]

From a historical-philosophical perspective, Comte establishes a three-stage law. Each individual science must go through these stages, just like all of humanity. The changing ideas of the sciences shape the entire rest of history. They show progress insofar as they move towards ever more adequate knowledge, which makes a return to an older and therefore lower level impossible. While the theological spirit and metaphysical thinking each only overlooked their own epochs, the positive spirit is able to grasp also the historical epochs of the past and their developments with their respective laws. "L'esprit positif, en vertu de sa nature éminemment relative, peut seul représenter convenablement toutes les grandes époques historiques comme autant de phases déterminées d'une même évolution fondamentale, où chacune résulte de la

---

[35] Eley 1998, 155.
[36] Jolibert 2004, 105–121.

précédente et prépare la suivante selon des lois invariables."[37] According to Comte, European history is thus divided into three epochs: a theological or fictitious, a metaphysical or abstract, and a scientific or positive stage. The latter is characterized by industrialization and a scientifically trained leadership class.

In an encyclopedic law, Comte establishes the hierarchy of basic sciences according to their objects and methods. The encyclopedic ranking among the sciences gives priority to the natural sciences and considers astronomy as the original source of the positive spirit. The regularity of astronomical phenomena, in fact, shows a real order that is completely independent of any human influence and thus conveys the basic feeling for the immutability of natural laws. The cultic worship of the stars has already led from fetishism to polytheism.[38]

However, mathematics is the starting point in the hierarchy, followed by astronomy, physics, chemistry, biology, and finally sociology.[39] These six disciplines can be grouped into pairs for simplicity's sake, according to Comte, resulting in a mathematical-astronomical starting pair, a biological-sociological ending pair, and a physical-chemical middle pair. Mathematics, with its branches of arithmetic, geometry, and mechanics, is at the beginning because it is the only necessary origin of rational positivity for both the individual and humanity. The basis is therefore the most general and abstract discipline, mathematics, while the endpoint is the most subjective and concrete discipline, sociology. Methodologically, mathematics uses logical proof, astronomy adds observation, physics adds experiment, and chemistry adds classification. Biology also proceeds comparatively and sociology historically. The order thus shows an increasing proximity to humans and a growing complexity, which is why positivist education should also be oriented towards this order.

Against Comte's three-stage law, the objection was raised that even today in the major European cities, a large number of people are still in the theological or metaphysical stage. However, this can be refuted by indicating that different stages can coexist. Max Scheler pointed out that religion is about personal salvation and metaphysics is about the essence and meaning of the world. These goals are not pursued by natural science, which investigates

---

[37] Comte 1844, 61.

[38] "Le sentiment fondamental de l'invariabilité des lois naturelles devait, en effet se développer d'abord envers les phénomènes les plus simples et les plus généraux, dont la régularité et la grandeur supérieures nous manifestent le seul ordre réel qui soit complètement indépendant de toute modification humaine. Avant même de comporter encore aucun caractère vraiment scientifique, cette classe de conceptions a surtout déterminé le passage décisif du fétichisme au polythéisme, partout résulté du culte des astres." Comte 1844, 107.

[39] "On parvient ainsi graduellement à découvrir l'invariable hiérarchie, à la fois historique et dogmatique, également scientifique et logique, des six sciences fondamentales, la mathématique, l'astronomie, la physique, la chimie, la biologie, et la sociologie, dont la première constitue nécessairement le point de départ exclusif et la dernière le seul but essentiel de toute la philosophie positive." Comte 1844, 101.

the causal relationships in the interest of dominating nature. With Comte, however, natural science is absolutized and turned into a pseudoscientific worldview.[40]

Like Bacon, Comte also gives priority to the natural sciences. The clarity and precision of mathematics, which, unlike theology and metaphysics, does not ask about goals, provides the structures for observations in astronomy, physics, and chemistry. Sociology becomes social physics and psychology becomes phrenological physiology. When Comte labels facts as "positive", he means not only the actual as opposed to the imagined. He also refers to what corresponds to common sense, foresight, and evaluatively to the contrast to the negative. Theological spirit and metaphysical thinking belong to the past according to Comte's three-stage law. However, because Comte considers religion to be an indispensable social bond, he places humanity with its culture at the center of his new religion, establishes a canon of good books as a new censor, and replaces the saints in his positivist calendar with different fields of knowledge and venerates their most outstanding representatives. At this point at the latest, the question arises whether thereby he does not fall back into the stages of theology and metaphysics. This did not detract from his impact as the main representative of French positivism.

6.5 Causality

In antiquity, Empedocles, Democritus, and Epicurus argue against natural goals and purposes. Lucretius, for example, argues that the eye was not created to see, or the tongue to speak, but rather the existing faculties form their use. In contrast, Lactantius argues that fish do not swim by chance and humans do not think by chance, but each living being must serve the purpose for which it was created.[41] In modern times, the arguments against teleology increase. According to Descartes, the nature of a thing is not at the same time its purpose. To attribute purposes and intentions to God would also be an impermissible anthropomorphism, attributing human characteristics to him. Spinoza goes even further when he defines purposeful action as striving for something that the striver needs. Applied to God, such striving would mean the loss of perfection. In addition, like Descartes, Spinoza rejects attributing human action towards a purpose to all natural things by analogy. In reality, teleological explanations of processes replace unrecognized causalities and are results of the imagination.

Paul Janet, who taught at the *Faculté des Lettres de Paris*, published a 748-page standard work on the topic of final causes in 1876. Right at the beginning, he defines the final cause under reference to Aristotle as a cause

---

[40] Fetscher 1994, XXXIII–XXXV); there were precursors: Lenoble speaks of a kind of positivism in Mersenneand Gassendi in contrast to the metaphysics of Descartes: Lenoble 1943.

[41] Hoffmann 2004, 1492.

in which an action is performed to achieve a goal. If, for example, the goal, health, is the cause of a walk, then depending on the perspective, health can be the cause or the effect. It can thus be considered as a "cause de sa propre cause"[42] , as the idea of an intended effect. If the eye is the cause of vision, then vision is the final cause of the eye. The chain of final causes is thus the reversal of the identical chain of effective causes. The question now is whether the final cause belongs to such general principles as identity, causality, substance, or space and time. If it is true that nothing happens without a cause, can one also assert that nothing happens without a goal, or according to Aristotle "La nature ne fait rien en vain"?[43] Why was one stone thrown to the left and another to the right during a volcanic eruption? Is the final cause more likely to be located in the area of organic life and psychology, while the effective cause belongs to the mechanical world?[44]

In this context, since the words goal and purpose have the same meaning, we also use them synonymously to denote a final cause or for teleological thinking. Generally, in the case of the final cause, there seems to be a transfer of human purpose-means terminology to other areas, e.g., to a universal purpose or meaning event. Already Bacon and Spinoza opposed nature-immanent purpose causes, while Johann Wolfgang von Goethe notes: "When the teleological explanation was banished, nature was deprived of mind; there was no courage to attribute reason to it, and it finally lay spiritless."[45] Of course, the situation from the perspective of Historical Materialism with Karl Marx and Friedrich Engels looks quite different, as their societal structure is determined by what and how is produced and how the produced is exchanged. This leads to the fact that causes are not to be found in the minds of people, but in changes in the ways of production and exchange. Therefore, Engels welcomes that Darwin's theory of evolution has destroyed teleology.[46] Nature does not proceed purposefully, but through enormous waste of life germs, with the failure of the initiated being the rule and success the exception. The latter is admired by teleology too shortsightedly. The mechanism of adaptation becomes a new kind of purposefulness. The religious consequences of turning away from teleology become clear with the physician and biologist Ernst Haeckel when he demands the replacement of religion by Darwinian evolutionism.[47] From his point of view, man as a natural product of evolution has as many rights as any other product of evolution, e.g., as a cobblestone. Dignity and respect are only human inventions. Morality and

---

[42] Janet 1876, 2.

[43] Janet 1876, 6.

[44] Janet 1876, 191; see also Eisler 1914.

[45] Goethe 1949, 464.

[46] Engels to Marx 11./12. 12. 1859; MEW 29, 524; cit. after Schlüter 1998, 974.

[47] Löw 1994, 118; cf. also Spaemann, 1986.

beauty are results of the Big Bang, matter, and natural laws and serve at best as survival advantages with an illusory character.

The positivism that emerged in France in the first half of the 19th century has lasting consequences for teleology. It began to take shape in the Enlightenment of the 18th century, when the usefulness of the mechanical arts became the guiding principle and the *vita comtemplativa* became secondary to the *vita activa*.[48] Auguste Comte (1798–1857), building on this, sees in his three-stage law the positivist science as the third and highest form of natural explanation. For him, mathematics as a basic science with its areas of arithmetic, geometry, and mechanics is at the top. The model becomes analytical mechanics, which dispenses with metaphysical speculations and ontological assumptions.[49] Of course, opposition also forms against this.

To the opponents of positivism, a restriction to experimental science appears to be associated with the loss of important areas. They therefore compare it to being on an island surrounded by an ocean without a ship and sail. Such a narrowing would lead to the loss of ethical principles, metaphysics, and religion. Overcoming this is the goal of Auguste-Théodore-Paul de Broglie (1834–1895), who sees himself as a professor of theology and the spokesperson of a "mouvement néochrétien".[50] Again, it is the doctrine of final causes that provides an important argument. "Toute plante, tout animal se développe et grandit conformément à une idée directrice."[51] If an engineer has assigned a motor and transmission systems to a locomotive for the purpose of transporting people or goods, he had this goal in mind when he started work. According to de Broglie, this can be compared to the body of a horse, where the organs and muscles also serve a purpose. Why should there not also have been a guiding intelligence at work in the works of nature? The fact that the future determines the present in the final cause is shown by the example of a child, in whom the eyes are formed before birth so that it can see after birth.[52] The self-imposed limitations of positivism thus prevent knowledge according to de Broglie.

What history did purposeful thinking and teleology have before they became the subject of controversy in the 19th century? The Greek word τέλος, *telos,* means fulfillment or goal, so it has something normative. The corresponding Latin *finis* is initially thought of as a boundary marker. The underlying idea is that everything has a well-defined place, its boundaries, but also its purpose of existence. To the purposeful world, Plato opposes a divine

---

[48] Rivero 2020, 263–264. If the idea of utility then also leads to instrumental reason in the sense of Horkheimer and Adorno, then its problematic nature becomes apparent: Rivero 2020, 120.

[49] Pulte 2015, 508.

[50] Broglie 1894, 3.

[51] Broglie 1894, 164.

[52] Broglie 1894, 173.

world builder who wants everything to be good. Because the philosophers know and pursue the purpose of life, they should rule in the state. Plants do not root in the air, but in the earth with the goal of protecting themselves. An inner tendency towards one's own perfection causes the goal-oriented activity, through which the respective definiteness and essential form is achieved. In ethics, happiness is the right goal, which is achieved through virtue.

When physicist Isaac Newton (1643–1727) believes that the natural scientist can recognize the wisdom of the creator precisely in the final causes, Gottfried Wilhelm Leibniz (1646–1716) adds that in physics, mechanical causal causes always also depend on final causes, since there is nothing but activities, every activity is self-activity, and thus all reality is to be seen as purposeful activity of the involved things (monads). Building on this, Christian Wolff (1679–1754) criticizes Spinoza and counters him that God wants nothing without a purpose, and that the world is the best of all possible worlds, which he created with the goal of revealing his greatness and goodness. The final causes were thus a much-discussed topic from antiquity to the 18th century. What becomes of this in the 19th century?

In France, the developmental biologist Jean Baptiste de Lamarck (1744–1829) had asserted in his 1809 published *Philosophie zoologique* that changes in environmental conditions, i.e., the milieu, cause changes in organisms and create new organs, which are further developed through use, which is also established in the genetic material over generations. The current of Neolamarckism, on the other hand, emphasizes the ability of organisms to adapt to given conditions and to inherit acquired abilities, which was later seen as a supplement to Darwinian ideas, which trace evolution back to natural selection of hereditary variants. According to Charles Darwin's widely disseminated theory of evolution since 1859, only organisms that have adapted their properties to the environment survive the struggle for existence. Thus, the teleology of organisms is replaced by the random combination of mechanical causes. Darwin is supported by the triumph of the natural sciences in the 19th century.

This is opposed by vitalism, which Leibnizhad previously defended against Hobbes, asserting that mechanical laws cannot produce an organism, but rather a life force is needed. A prominent representative is the French physician Claude Bernard (1813–1878), who sees the uniqueness of the living cell in a *milieu interieur*. This stands in contrast to a *milieu exterieur*, i.e., the outside world. With the higher degree of organization of the living being, its independence increases.[53] The consideration of the interrelationships between living beings and their environment was seen by Isidore Geoffroy Saint Hilaire (1805–1861) as the task of ethology, a kind of behavioral biology.[54]

---

[53] Bernard 1865, 109.

[54] Saint Hilaire 1854.

Are there atoms that live? The question of infinitely small microorganisms occupies Michelet in the second half of the 19th century in his considerations on natural history. Preceding this in the first half were Lamarck's transformation theory, the unitary composition plan in nature represented by Geoffroy Saint-Hilaire, and Cuvier's concept starting from a lack of uniformity. Michelet's *La Mer* opposes a divine creation and instead vitally assumes a material cause as an eternal source with spontaneous and permanent reproduction, drawing on Lamarck's *effort*, i.e., striving and force. This cause is for him a will to progress and a constant urge for improvement, which he also wants to transfer to the social and political level. Republican and scientific ideas are to replace Christian notions.[55]

The main theses of Darwinism are briefly presented below from the perspective of his opponents. The archiepiscopal seminary prefect in Freising Franz L. Grassmann won a prize by answering a question posed in 1884/5 by the theological faculty of the University of Munich comparing Augustine and Darwin in his writing. As a precursor to Darwin, he mentions Buffon, who originally assumes a certain number of basic types in the animal kingdom, from which the existing species descend by degeneration, and Lamarck, for whom the present animal life developed from worms and small animals like infusoria due to the causes of external circumstances, needs, and habits. Then he summarizes Darwin's hypotheses, who in turn assumes original forms for the plant kingdom and the animal kingdom. The breeding of domestic animals, where the breeder selects the best specimens, can serve as a comparison. Although all pigeon breeds descend from the blue rock pigeon, if one is unaware of this origin, one might consider all varieties as distinct species. What happens through domestication happens in nature by itself. The struggle for existence is won by the organisms that are advantageously modified. These advantageous modifications are passed on through inheritance. This principle, by which every slight, if only useful, modification is preserved, is called natural selection.[56] Therefore, harmful variations perish. The prerequisite is the tendency of organisms towards directional and endless variability.

Natural selection is complemented by sexual selection, through which the most beautiful and strongest prevail in the competition and ultimately have the most offspring. The principle of correlative change ensures that when one part of an organism is modified, other parts are also modified. The principle of use and disuse, on the other hand, explains why unused parts of an organism atrophy, as can be observed in the eyes of moles. External living conditions according to Darwin are less important than the tendency of organisms to vary. In order for an ape-like creature to be transformed into a human, it is necessary that this earlier form, as well as numerous subsequent links, have all

[55] Séginger 2019, 31–48.
[56] Grassmann 1889, 93.

varied in mind and body.[57] However, since it was advantageous for the evolving human to have the arms and upper body free, only those individuals who were modified in this way would have survived in the struggle for existence. Why did man lose the hair of the ape? Here, only sexual selection could have been active. First, the female sex gradually lost it, and through inheritance, the male sex was also freed from it.[58] The development of human intellectual abilities is the result of the continued use of language, with the perfection of the speech organs affecting language ability. Individual virtues do not result from natural selection, but from habit, instruction, and example.

Darwin draws on the similarity of construction plans to support his theses. Thus, the skeletal structures in the hand of a human, in the wing of a bat, and in the leg of a horse are similar. Grassmann critically argues that no transformations of species are occurring in the present time and there is no evidence that coral animals have changed in the last 30,000 years.[59] With his summary of Darwin's theses, Grassmann aligns himself with Darwin's opponents. Since these prefer to orient themselves to the medieval philosophy of Thomas Aquinas, his understanding of the category of the final cause will be demonstrated in the following.

Since Thomas Aquinas has shaped medieval scholasticism and is therefore a reference philosopher for neo-scholasticism in the 19th century, he is now presented in more detail. Thomas Aquinas sees nature and man oriented towards a highest purpose. "The purpose is the general category under which the entire universe as well as the life of the individual is considered."[60] Since Thomas follows Aristotle, he also sees the act of will directed towards that goal which Aristotle defines as "the 'what for'". One takes a walk to become healthy. The house that the architect wants to build is the goal of his will activity, to which he subordinates other things.[61]

Even natural beings without reason, according to Thomas, have a natural striving towards a goal, an *appetitus naturalis*. What is will in humans is inherent in animals in sensuality. The main goal is self-preservation, which is linked to self-promotion. By preserving the species, nature participates in a kind of immortality. In humans, reason and will dominate sensuality. Perfect good and therefore the goal of man is for Thomas and Aristotle the felicity. Every being carries an immanent purpose within itself: it should perfect itself and bring the possibilities inherent in it to full development. The means must be in the right proportion to the goal. Since reason is the essential

---

[57] Grassmann 1889, 102.

[58] Grassmann 1889, 105.

[59] Grassmann 1889, 116–117.

[60] Steinbüchel 1912, 1.

[61] See chap. 3.3 Teleology.

characteristic of man and every action should be in accordance with the form of being, actions contrary to reason are unnatural in him. Since, according to Thomas, there can be no higher thought than the knowledge of God, this is the most perfect activity of man.[62]

The concept of the goal has consequences for the conception of society and the cosmos. Man as a social being is dependent on others. The basic condition is the subordination of the many to a ruler who has the common good in the state as his goal.[63] Other purposes of the state are the maintenance of order internally and externally, the material well-being of its citizens, and the cultivation of intellectual and ethical goods. When Thomas conceives the universe teleologically, man is the culmination point, as the less developed always serves a higher goal. The parts of the universe exist because of the whole. The world appears as an ordered army, over which the commander, i.e., God, stands as the ultimate goal and highest good. Therefore, creatures strive for similarity with him through imitation. The ultimate goal of all things is to become similar to God. According to Thomas, the order and purposefulness of the world can be used to prove the existence of God. The norm of ethical action is shown to man in a natural law, a *lex aeterna,* as a divine world plan.[64] Evil in this context is only a lack of goodness. It becomes apparent that for Thomas Aquinas the doctrine of the goal and purpose is central, not only for irrational nature and rational man, but also for the state and the cosmos. He thus provides the basic assumptions that could be taken up at the end of the 19th century to counter the new worldview shaped by causal explanation.

In the following, two journal articles will be presented that, with reference to Thomas Aquinas, develop a neo-scholastic theory. The *Revue Thomiste. Questions du temps présent,* founded in 1893, aims to address current social and philosophical issues and to help science remain or become Christian, help scholars remain or become believers.[65] Starting from Thomas, the compatibility of faith and knowledge, of the Bible and science, is to be demonstrated. The fact that the first article deals with the teachings of evolution and is continued in several installments from 1893–1896 shows the great importance attached to this topic. The Thomist Ambroise Gardeil (1859–1931) refers to Darwin's theory that new species have emerged due to changing conditions of existence as transformation or evolution theory and sees it as generally accepted in his time as the theory of gravity. It is also applied in various fields such as geology, biology, cosmology, psychology, and sociology. Since it starts from matter and not from spirit, it is materialistic and not spiritualistic.

---

[62] Steinbüchel 1912, 92.

[63] Steinbüchel 1912, 104.

[64] Steinbüchel 1912, 126–147.

[65] Gardeil 1893b, 2.

In order to relate the theories of Thomas to contemporary teachings, Gardeil first explains how Thomas dealt with antiquity in his treatise *De distinctione rerum*. The pre-Socratic philosopher Democritus, already mentioned, saw only material causes in natural events, while the pre-Socratic philosopher Anaxagoras made an acting and forming spirit the basis. While the former represents a purely chance-determined materialism, the latter advocates teleological thinking. Thomas summarizes the former doctrine in the sentence "Forma est propter materiam," the latter in the sentence "Materia es propter formam," whereby form, goal, and purpose again mean the same: "Finis et forma coincidunt."[66] For the teleological thinking of Thomas, an oak remains an oak, from the acorn to the mature tree, while—so Gardeil thinks—the random, lawless movements of Democritus' atoms can hardly explain how individual parts of matter become a constant unit. Gardeil cites Ernst Haeckel as a follower of Democritus' teachings and as a disseminator of Darwin's teachings. His mechanistic doctrine has the advantage of being as clear as it is striking and thus attackable.[67]

The clear alternative between materialistic evolutionary theory and teleology, which Thomas still saw when considering their ancient representatives, had become complicated by new arguments by the 19th century. In teleology, a distinction has been introduced between an immanent and a transcendent type. The oak tree could be imagined as an example of the former, and the house, where the architect's plan is outside the house, as an example of the latter. In the sense of Leibniz, one could assume that the world was created by a transcendent act, but then left to itself and to immanent goals. From a Thomistic perspective, Gardeil considers this acceptable and recommends the assumption of an external mover and of original forms of nature that determine the direction of development.[68]

The doctrine of the unconscious introduces an irrational element into theories of action. Here Gardeil deals with the *Philosophie des Unbewußten* (1869) of Eduard von Hartmann, who influenced Sigmund Freud and Carl Gustav Jung with his doctrine. An action that is determined by the unconscious can neither be purely materially caused nor consciously derived from a purpose or goal. It is therefore neither simply causal nor finally conditioned. Unconsciously, for example, the sheep flees at the sight of the wolf. However, Gardeil sees insurmountable difficulties when the unconscious is related to the evolutionary process and new species spontaneously emerge from the unconscious, from an "idée-volonté inconsciente et immatérielle [...] immanente et unique."[69]

---

[66] Gardeil 1893a, 37.

[67] Gardeil 1893a, 727.

[68] Gardeil 1893a, 327.

[69] Gardeil 1895, 64.

Can a change of species also take place within the framework of teleological thinking? Here Gardeil introduces the Aristotelian concept of *habitus*, or habit, from the second book of Ethics and quotes its commentary by
Thomas. Only through action can one acquire virtue, just as one can only
learn something in the arts through practice. Habits and attitudes are formed
through practice.[70] Gardeil illustrates this with the hand of the pianist,
which changes muscles and nerves through constant practice in the interest
of the goal of perfection. Changes should be imagined in the same way in
other areas. Just as repetition in humans brings about a change, so it happens in the whole world. This makes habitus the basic principle of a goal-oriented and Thomistically compatible understanding of evolution.[71] It should
be emphasized that here the goal of the respective organism is consciously
set. Through repetition, an evolution of the original orientation towards a
new one is created.[72] Since, according to Thomas, both man and the whole
world are determined by reason, namely by human or divine reason, Gardeil
opposes Darwin's theory of evolution with an *évolutionisme des habitudes* and
concludes that species are to be distinguished from each other by habits and
attitudes: "Les espèces sont pour nous des habitudes de la matière."[73] He
believes that he has thus contributed to the reconciliation of faith and knowledge, of scholasticism and natural science.

Gardeil justifies exactly what his also Thomistically influenced colleague,
Vincent Ermoni (1858–1910), vehemently rejects as Darwin's theses, namely
that organs arise through habits and exercise: "L'organe s'acquiert par l'hab-
itude, par l'exercice." And: "La fonction crée l'organe, elle en est le principe,
loin d'en être le résultat."[74] In Ermoni's article *Finalisme et antifinalisme*,
which appears in several installments of the *Annales de philosophie chrétienne*
from 1892–1893, finalism stands for Thomistic teleology and antifinalism for
Darwinism.

Ermoni focuses on the question of whether the effect is determined by a
goal or by practice and habit. For Thomas, the purpose of the eye is to see,
which is also its effect. In contrast, the Darwinists now claim that birds fly
because they have wings. And in response to the question of how such a perfect construction as the wing of a bird could have arisen, they cite practice,
constant use, which has refined and produced the organ. This, they claim, can
be compared to a misshapen knife that becomes sharper and easier to han-
dle through constant use. Therefore, for the Darwinist, practice and function
create the organ and it does not precede functioning. Only after the function

[70] Gardeil 1896, 75, 77.

[71] Gardeil 1896, 85.

[72] Gardeil 1896, 225.

[73] Gardeil 1896, 245.

[74] Ermoni 1892, 347.

has created the organ can it become active.[75] Polemically, Ermoni counters with the question of how exactly it is to be explained that the fins of fish have become the wings of birds. Where did the sudden impulse come from to use fins as wings? Were the environmental conditions unfavorable for fish?

In teleology, according to Ermoni called *finalisme*, it is said that the organs serve the purpose of their functions. He follows Thomas and Aristotle, in whom—as already mentioned—the formal and the final cause often coincide, when for him, seen teleologically, even the parts of a whole serve this whole, which is their purpose. In a "solidarité morale" or a "solidarité physique"[76] the parts produce a result and are subordinate to the whole.

Ermoni therefore considers a focus on meaning and purpose to be necessary. Therefore, a positivist view, which only records facts, cannot lead to science. The latter indeed starts from facts, but then needs general laws, which are impossible without finality: "S'il n'y a pas de finalité pour canaliser leur tourbillon inconstant et fugitif."[77] The stability of things is also not explainable without a purpose orientation. If the eye did not have the purpose of seeing, how could the physiologist assert with some certainty that the eyes of people born next year will be similar to the eyes of those currently living? From this it can be concluded that finalism does not hinder the progress of science, but promotes it.[78]

If one now accepts the necessity of a purpose, the question already posed by Gardeil arises again, whether the purpose lies outside or within the respective thing. If one relates the question to the universal purpose, the choice between atheism or theism depends on the decision for immanence or transcendence. As a solution, Ermoni suggests distinguishing between the goal to be achieved and the tendency inherent in the thing. Then a purpose could be both immanent in terms of the tendency and transcendent in terms of the goal. As an example, he cites the arrow with which William Tell splits the apple on his son's head. From the outside, it is the hand and eye coordination of the shooter, his experience, and the crossbow that guide the arrow to its target. But the arrow also has an inner tendency. It is shaped in such a way that air resistance is low and accuracy is high. Can the distinction between immanent tendency and transcendent goal be transferred to Darwin's theory of the adaptation of species in the animal kingdom to the environment? Is the length of the giraffe's neck the result of an external environment where the leaves can only be found at the top of the trees? Is the anatomical adaptation to the environment only conditioned by the environment, or is it not rather

---

[75] "l'exercice crée l'organe, qui par là même ne préexiste nullement à la fonction, mais lui est postérieur. [...] l'exercice serait cause, puisqu'il aurait créé l'organe; et, après cette création, il deviendrait effet, puisque l'organe entre en jeu et produit les actes." Ermoni 1892, 361.

[76] Ermoni 1892, 355.

[77] Ermoni 1892, 454.

[78] "Concluons donc: une doctrine finaliste, loin d'être une entrave aux progrès de la science, est, au contraire, de la plus grande utilité." Ermoni 1892, 455.

the maintenance of the organism, which would perish without this adaptation, the goal? If one accepts this goal as a purpose, then Darwinism can also be attributed a teleological component. Ermoni sees this line of thought already laid out in the scholastic thesis "operari sequitur esse", according to which activity is closely connected with being.[79]

One might now think that Ermoni is trying, like Gardeil, to reconcile Darwinism with scholastic terminology. But he draws different conclusions. For it is not by chance that the goals are achieved, but by external guidance.[80]

What consequences does the abolition of teleology have for the image of man? To what extent can the difference between man and animal be maintained? This question is addressed by the Jesuit Joseph de Bonniot in his book *La bête comparée à l'homme,* which appeared in its second edition in 1889. In it, he deals with the teachings of the naturalists of his time, especially the followers of Darwin, for whom man is a thinking animal and the animal is a man with intellectual aspirations.[81] This is the case when man is defined as "un animal vertébré, mammifère, de la classe des quadrumanes et de l'ordre des primates",[82] that is, as a vertebrate and mammal, equipped with two hands and two feet and belonging to the group of primates. With his skeleton, he is comparable to the monkey, the eagle, and the frog. Like all other vertebrates, he is equipped with a heart, lungs, eyes, a liver, a stomach, blood, nerves, and muscles. The difference between a man and a dog is less than that between a dog and a crocodile. Anatomically speaking, man is thus an animal. If one refers to intelligence, then the naturalists do not count the quality, but the quantity: Since the brain of the animal is smaller than that of man, it also has less intelligence. Bonniot counters this with the four areas in nature: the motionless mineral, the living plant, the animal endowed with sensations, and the human area endowed with reason, whereby reason is not a mere function, but a real substance. If the naturalists now claim that one has come from one area to another through evolution and transformation, Bonniot considers this a mistake.[83]

---

[79] Ermoni 1892, 501; Thomas also formulated it similarly when he saw things oriented towards a goal: "Cela ressort de ce que toujours, ou du moins plus fréquemment, elles agissent de la même manière pour atteindre ce qui est le meilleur." Ermoni 1892, 131.

[80] "d'un être connaissant et intelligent, comme la flèche est dirigée au but par celui qui la lance. Il existe donc un Être intelligent qui ordonne toutes choses à leur fin; et cet Être, nous l'appelons Dieu." Ermoni 1892, 131.

[81] He rejects the doctrine, "qui veut faire de l'animal un homme, afin de pouvoir faire de l'homme un animal." Bonniot 1889, X.

[82] Bonniot 1889, 30.

[83] "Il est absolument contraire à tous les principes qu'une réalité nouvelle passe de la non-existence dans l'existence par transformation; le néant transformé ne donne que le néant. Pour produire un être réellement nouveau, il faut nécessairement l'intervention d'un acte créateur." Bonniot 1889, 64.

Darwin also attributed language, a sense of aesthetics, self-awareness, thinking, religiosity, and morality to animals. For aesthetics, he cited the song of birds and the male display of colorful plumage in front of female birds, for religious feeling a dog that regards its master as a god, or the monkey that idolizes its keeper. Now, from a Darwinian perspective, if thinking is a function of the brain, just as digestion is a function of the stomach, and there is more thinking the more brain there is, then the ant is a beautiful counter-example. Assuming that the human brain weighs about 1,350 grams, then the weight of the ant's brain is 4 milligrams, which would make it 337,500 times less rational than a human. And do not the elephant and the whale have larger brains than humans, without being more rational, asks de Bonniot, to lead Darwinism *ad absurdum*.[84]

One could now dismiss such discussions from the end of the 19th century as *curiosa* of the past and claim that we have made much progress since then and that teleology has become redundant. Did the problem regarding the universe and society really no longer arise in the 20th century? To answer this question, let us first look at the contributions of a discussion at the *Deutsche Akademie der Naturforscher Leopoldina* of the former German Democratic Republic, which took place in Halle an der Saale in 1975.[85] The contrast between evolutionary theory and teleology was discussed here with reference to the opposition of chance and necessity.

The necessity of goal orientation is countered by theses such as that initial distributions are random, as are mutations and their order, and that fluctuations do not regulate themselves, but rather, they intensify. If one compresses the age of the earth of five or six billion years into a single year, man appeared on earth 30 minutes before midnight on New Year's Eve. This leads the discussion round to the following unresolved questions: Why has brain content increased in such a short time? Was it hunting, which required tools, communication, and organization? Or did the killing of other animals serve more as a show of strength in chimpanzees? And why did gorillas and chimpanzees not become human?[86]

Of the two areas of biology, the one that deals with reading genetic programs and decoding them cannot provide an answer; rather, it is the other branch of biology that deals with the emergence of these genetic programs. The former is limited to the question of how a process takes place, which it answers chemically and physically; the latter asks why, e.g.: Why did man come into existence, and why did he not emerge earlier? Was it chance or necessity?

At this point, Laplace is quoted, who thought that if he could describe everything that is going on in the world right now, then he could make

---

[84] Bonniot 1889, 228, 234, 316.

[85] Bruns 1975, 395–416.

[86] Bruns 1975, 411–414.

predictions far into the future—an idea that is immediately refuted with an example. If one looks at the animal world from 200 million years ago, it would be impossible to say which reptiles or mammals would rule the world 200 million years later. Primates have developed a wealth of species, all of which are becoming extinct, except for the one with a developed brain. This leads the discussion round to astonishing arguments in favor of necessity and goal: "The fact that evolution leads to more differentiated species up to humans and is by no means random cannot be denied." Or "Evolution must have been directed to such an extent that a breakdown could no longer occur."[87] What kind of goal and direction could be meant was not discussed in more detail and should also remain open here.

However, not only in evolution does the opposition of goal and effective cause, of necessity and chance, play a role. The model chosen is also not insignificant for the shaping of society. It is therefore not surprising that in the 1950s Thomas Aquinas and Plato became reference figures again. With both, the German-American critic of empirical philosophy Leo Strauss (1899–1973) legitimized the idea of natural law against a general historicization and relativization. From natural law he derived: All natural beings have a natural goal, a natural destiny, which decides what is good for them to do.[88] With this, he criticizes the social science of his time, which wanted to prove by referring to history that principles are changeable. Thus, Strauss concludes that historicism and positivism want to replace metaphysics and theology, thereby positioning himself against Karl Popper, who in 1945 saw Plato as the enemy of the open society.[89]

For Strauss, principles are unchangeable and thus not subject to historical change and chance. If principles were justified purely because a single society accepted them, then cannibalism would also have to be legitimized. But whenever unjust laws are spoken of, a standard beyond positive law is assumed. Since the explanation from desires and drives is not sufficient, the science of man must therefore be teleological. Plato distinguishes between the good and the ancestral. The question of the first things presupposes that they always exist in the same and imperishable form. Nature may be hidden by authoritative decisions, but it is older than any tradition. The good life has always been the life in accordance with nature.[90]

Strauss names Hobbes as the creator of political hedonism and atheism, who conceives man as an asocial being. As the founder of liberalism, he made human rights, not duties, the basic political facts. For the sovereign, whom Hobbes wants to install by contract, laws apply, which the people have to

[87] Bruns 1975, 410, 408.

[88] Strauss 1956, 8.

[89] Popper 1992.

[90] Strauss 1956, 88, 91, 94, 131.

submit to, by authority, not by truth or reason. Strauss opposes this, referring to Rousseau, who sees a radical error in it and proposes a departure from a world of sciences, artificiality, and conventionality and a return to the natural state.[91] Strauss sees himself on Rousseau's side. He criticizes Hobbes for turning away from natural law and thus relativizing it through contract agreements. Strauss rejects the latter as well as a world of historicism and positivism, where historical developments appear random and the view for necessary natural goals and determinations is obscured.

So, if in the 20th century, as with Strauss, a teleologically understood natural law is still contrasted with the relativizations of liberalism, historicism, and positivism, and as with the German National Academy of Sciences Leopoldina, the decision between necessity and chance is still open, but the orientation towards a goal is brought into play, then attempts at reconciliation between teleology and causality thinking do not seem to have prevailed. After all, Gardeil, having seen the controversy between Democritus and Anaxagoras summarized in the theses of Thomas Aquinas "Forma est propter materiam" and "Materia est propter formam", distinguished an immanent from a transcendent purpose and introduced the unconscious as a further problem, and proposed the Thomistic-Aristotelian concept of habitus to accommodate Darwinism with an *évolutionisme des habitudes*. In contrast, Ermoni found it unthinkable that practice and exercise could create new organs and turn the fins of fish into the wings of birds. He explains the stability of species by their purpose orientation, although he admits that one could also attribute a teleological component to Darwinism if one attributes an immanent component to the purpose in terms of tendency and determines the external component of the goal as transcendental. The impossibility of something new arising from transformation from the non-existent, and the absurdity of inferring thought from the quantity of brain mass, according to Bonniot, speak for the stability of species and the fundamental difference between man and animal.

It has thus been shown that the controversy over goal thinking and causality has a long tradition. While Empedocles, Democritus, Epicurus, Lucretius, Descartes, Spinoza, Comte, Bacon, Spinoza, Haeckel, Engels, Lamark, and Darwin turned against goal thinking, Plato, Aristotle, Lactantius, Thomas Aquinas, Newton, Leibniz, Wolff, Goethe, Bernard, and Saint-Hilaire spoke out in favor. The equally long and prominent chain of participants in this ongoing discussion demonstrates its importance for understanding man, society, and the universe. What is new in the 19th century is that the level of speculation is abandoned and the decision in favor of causality is made based on the natural sciences.

In summary, it can be stated that the English empiricists spread the method of inductively starting from experiences and experiments. Concepts

---

[91] Strauss 1956, 175, 188, 193, 263–265.

are no longer derived from substances, but obtained by generalization. The assumed uniformity allows future predictions. Ethics, politics, or theology should no longer be at the center of interest, but scientific knowledge that serves comfort and the mastery of nature. Inductive method and observation are also applicable to ethics and politics. The clarity and precision of mathematics, which does not ask for goals like theology and metaphysics, provides the basis for the individual natural sciences, which deal with experiences and positive facts, i.e., with the actual and not with the imagined. The rivalry between teleology and causality thinking is decided in favor of causality, despite intense controversies and repeatedly raised counterarguments, not least because the dominant natural sciences work with cause research. Thus, there is a paradigm shift from the subject to the object, from dealing with human action in ethics and politics to scientific law, from teleology to causality, from deduction to induction, from the priority of mental concepts or substances to positive facts. The question is no longer what a thing actually is, but how it works and can be made useful. How literature reacts to such changes will be illustrated by a few examples in the following.

6.6   Novel as Experiment

In his treatise *Le roman expérimental*, the naturalistic French novelist Émile Zola (1840–1902) aims to transfer and adapt the experimental method, as developed by Claude Bernard in his *Introduction à l'étude de la médecine expérimentale* for medicine, to the field of literature. The physiologist Claude Bernard (1813–1878) makes a clear distinction between the observer and the experimenter. While the former observes the phenomena of nature that he cannot change, the latter modifies and changes them to analyze them under conditions and circumstances that are not naturally given. A typical example of an observer is the astronomer, who cannot vary the celestial bodies through new experimental arrangements.[92]

If experiments in the physical realm yield insights, then according to Zola this also applies to the human realm of "de la vie passionelle et intellectuelle".[93] In the preface to the novel *Thérèse Raquin*, Zola explains that he deliberately chose characters whose actions are determined by physical and not by intellectual factors. Each of their actions is physically controlled by blood and nerves and not by free will, with the soul being completely absent. Therefore, it is important to find out the conditions that determine the occurrence of phenomena. If physical bodies are determined by causes, then this also applies to living beings. After initially having experimental physics and experimental chemistry, we will soon have experimental physiology and the experimental novel. Once it has been proven that the human body

---

[92] Bernard 1966, 27.
[93] Zola 1971, 60.

is nothing more than a machine, the experimenter can recognize its workings. It should be possible to trace the processes in the brain back to scientific laws. Only in this way will the imaginative novels of the idealists be replaced by novels with observation and experiment, the *romans d'observation et d'expérimentation*. Bernard also sees an increase in the degree of complexity from physics to chemistry and biology through to physiology and medicine. He suggests breaking down the complexity of phenomena into the simplest possible individual parts through analysis in order to experiment with them more effectively: "Ils ramèneront ainsi les phénomènes à leurs conditions matérielles les plus simples possible, et rendront ainsí l'application de la méthode expérimentale plus facile et plus sûre."[94]

While experience should be free of preconceived ideas, an experiment requires a hypothesis that anticipates the result of the experiment. The experiment now has the task of confirming or falsifying the hypothesis. The experiments are to be arranged in such a way that their results allow for the testing of the hypotheses.[95] If individual phenomena refute a hypothesis, then they drive the sciences forward to new hypotheses. The starting point and basis are real facts, from which the inventive power gains further phenomena to demonstrate the mechanisms at work. Scientists and novelists take the same approach when they want to have their hypotheses confirmed by reality: "Il se place devant la nature, a une idée a priori et travaille d'après cette idée."[96] The novel itself becomes an experimental protocol. Where there used to be descriptions in the novel, there is now an inventory of the milieu that complements and predetermines the human being.

For understanding, Bernard's somewhat more precise explanations of the hypothesis are presented. He sees this in the experiment as a vague hunch or intuitive assumption that the experimenter draws from within himself. Only afterwards does he approach reality to experimentally test the assumptions. Bernard uses the terminology of Kant when he refers to the hypothetical idea as a priori and its experimental verification as a posteriori.[97] Such a basic idea was also the starting point of scholastic metaphysics. However, this was considered absolute truth, from which conclusions were then drawn with logical consequence, while the more modest experimenter wants to confirm or refute it through his experiment. The basic idea, which is a hypothesis for the experimenter, arises according to Bernard from intuition and is further developed by reason, until it is put to the test through experience and experimentation.

---

[94] Bernard 1966, 85, 70, 71.

[95] "L'expérimentateur est celui qui, en vertu d'une interprétation plus ou moins probable, mais anticipée, des phénomènes observés, institue l'expérience de manière que, dans l'ordre logique des prévisions, elle fournisse un résultat qui serve de contrôle à l'hypothèse ou à l'idée préconcue." Zola 1971, 63.

[96] Zola 1971, 94.

[97] Bernard 1966, 38.

"Le sentiment engendre l'idée ou l'hypothèse expérimentale, c'est-à-dire l'interprétation anticipe des phénomènes de la nature."[98] As an example for illustration, the hypothesis arising from observation is cited that all swans are white. If the observation is continued, e.g., in other countries under different conditions, then the sighting of a single black swan can falsify and refute the hypothesis.

Citing Darwin , Zola emphasizes the importance of inheritance and the respective circumstances. The experimental novel is a consequence of the scientific development of the 19th century. For it replaces the previous abstract and metaphysical or scholastic and theological preoccupation with man, which led to classical and romantic literature, with the consideration of human nature under the influence of physical and chemical factors and the impact of the environment.[99] Natural sciences thus form the basis for a novel author. Not only does it make clear the functioning of man thought of as a machine, but with the laws then found, it can support political science and economics and serve the practical purpose of transforming society into a better state. The goal of science is also according to Claude Bernard to control or predict phenomena. Zola leaves the rage of Achilles and the love of Dido in antiquity and on beautiful images. Now he wants passions like rage and love to be analyzed experimentally, not philosophically. "Le point de vue est nouveau, il devient expérimental au lieu d'être philosophique."[100] Natural sciences thus form the basis for a novel author.

Zola's sense of progress is evident when he portrays his own age as one of transition from a detestable past to an unknown future. Not only in technology with inventions such as the railway, telegraphy, or steam navigation, but also in politics and religion, everything is in motion. Against this background, the education system also needs to be reformed. If naturalism leads to anti-clericalism and requires a new morality, then it points the way to a new way of life.[101]

Since inheritance becomes visible over long-term developments—in the case of the Rougon-Macquart family it spans five generations—Zola was also able to include the historical level, focusing on the *Second Empire*. For him, the period of Emperor Napoleon III's reign between 1852 and 1870 is the moment that determines people. Thus, he refers to the history of the Rougon-Macquart family in the subtitle as *Histoire naturelle et sociale d'une famille sous le Second Empire*.[102] In Zola's novels, no single figure is the hero, but a family, on which he exemplifies the biological laws of inheritance, which

---

[98] Bernard 1966, 44.

[99] Zola 1971, 74.

[100] Zola 1971, 97.

[101] Becker 2014, 9–33.

[102] Daus 1976, 43–44.

for him are on the same level as the physical laws of gravity. In the case of the Rougon-Macquart family, a growing degeneration becomes apparent with the progressive exhaustion of vitality. The family lineage, like that of the nobility, is strongly determined by the past. However, while the family tree of the nobility served to demonstrate power and social precedence, here it becomes an expression of hereditary defects and moral transgressions. Although the dominance of the biological stands above that of the social, the laws of the milieu, by which Zola means place of residence and profession, also react. For Hippolyte Taine, the three factors of descent, milieu, and historical moment, "la race, le milieu et le moment", were decisive, but for Zola, the inherited inner dispositions have priority.[103] The railway appears symptomatic of progress and modernity in the 19th century. It transforms steam power into mechanically produced uniform motion, impresses with its speed, which opens up new spaces, and is made possible by massive investments in machinery and the rail network. In Zola's novel *La Bête humaine* (1890), the railway operating between Paris and Le Havre is a central motif and symbol of progress. If the railway mechanic Jacques Lantier is portrayed in his relationship with a young woman as a pathological and instinct-driven primitive man, this does not contradict the scientific thinking, but confirms it, as according to Zola's theory of inheritance, Lantier's behavior is biologically determined.[104]

Zola, who believes in progress, aligns the novel with natural science by viewing the human body as a machine, with whose workings the experimenter can play. The mental life and feelings are then analyzed in a further step through experimental physics, chemistry, and physiology. Thus, the old hypotheses of philosophers and writers are overcome, and the experimental novel emerges. Zola formulates this as follows:

> "Quand on aura prouvé que le corps de l'homme est une machine, dont on pourra un jour démonter et remonter les rouages au gré de l'expérimentateur, il faudra bien passer aux actes passionnels et intellectuels de l'homme. Dès lors, nous entrerons dans le domaine qui, jusqu'à présent appartenait à la philosophie et à la littérature; ce sera la conquête décisive par la science des hypothèses des philosophes et des écrivains. On a la chimie et la physique expérimentales, on aura la physiologie expérimentale; plus tard encore on aura le roman expérimental."[105]

What does "experimenting" mean in Zola's experimental novel? In the social sciences and in the novel, it is certainly not the same as in physics, as here the space of the indeterminate and uncertain is larger. Zola sees a difference in

---

[103] Bender 2009, 201–206; on the problematic transfer of scientific models to literature: Hajduk 2005, 236–253.

[104] Bender 2011, 103–126.

[105] Zola 1971, 70.

complexity between the physical *milieu externe* and the vital *milieu interne*. Since no binding laws or final proofs are possible in the analysis of human passions and actions, only hypotheses can be formulated in the social sciences and in the experimental novel, which should be based on precise observations but leave room for experimental arrangements.[106] While Balzac is concerned with the entire society, Zola selects a family as a sub-area. The characters in Zola's works initially appear very different, until their similarity becomes apparent from the laws of inheritance.[107]

Zola's stories of the Rougon-Macquart family were published between 1871 and 1893. However, the understanding of inheritance changed during this time. While Lamarck still advocated in his 1809 published *Philosophie zoologique* the theory that traits acquired during a lifetime are inherited, this was already refuted by Mendel around 1865. Mendel's teaching that innate genes are inherited, but not acquired ones such as a propensity for criminality or alcoholism, only became known to a wider public through August Weismann in 1883. The Weismann barrier prevents the mixing of acquired cells with innate ones, thus invalidating Zola's starting point. Zola explicitly mentions Weismann's theory in *Le docteur Pascal,* the last work of the Rougon-Macquart cycle, outlines the history of science from Darwin via Haeckel to Galton, to finally suspect that the future belongs to Weismann's teaching.[108] In *Le docteur Pascal*, the supposed progress is also doubted by the character of Clotilde, when she sees the hopes placed in the natural sciences for insights into the secrets of the world deferred to an ever more distant future and the knowledge rather decreasing than increasing.[109]

Two basic attitudes are confronted in Zola's *Le docteur Pascal*. On the one hand, there is the scientist and doctor Pascal, who collects empirical data on the history of his family and stores it in a cupboard to study the laws of inheritance on his own family. On the other hand, there is his mother, who sees a threat in the possible insights into the bad sides, as she wants to preserve a beautified image of the family. When she destroys the documents after Pascal's death, illusion and imagination defeat science, which in Pascal's case is content to confirm or refute hypotheses through empirical facts. The fact that the laundry of the son fathered by Pascal is stored in the cupboard that previously held Pascal's scientific documents can be interpreted as a victory

---

[106] Ventarola 2010, 290.

[107] Klinkert 2010, 202.

[108] "Il était donc allé des gemmules de Darwin, de sa pangenèse, à la périgenese de Haeckel, en passant par les stripes de Galton.Puis, il avait eu l'intuition de la théorie que Weismann devait faire triompher plus tard, il s'était arrêté à l'idée d'une substanceextrèmement fine et complexe, le plasma germinatif, dont une partie reste toujours en reserve dans chaque nouvel être, pourqu'elle soit ainsi transmise, invariable, immutable, de géneration en géneration." Zola 1893, 38; cf. also Popowicz 2019, 159–170; Séginger 2020, 88.

[109] Zola 1893, 94.

of life over science.[110] If Pascal could not apply his scientific knowledge and finally dies of arteriosclerosis, then the novel does not celebrate positivist natural science, but shows its limits.[111]

Zola's experimental novel is conceived as the result of the transfer of physical, chemical, and physiological methods to literature. In this, he sees humans physically determined by nerves and blood, taking into account neither free will nor psychological factors. Literature is experimental when it—like the physicist—verifies or falsifies a hypothesis through experiment. This way, phenomena can be controlled or predicted. Since in Zola's work inheritance plays a larger role than the social milieu, the Weismann barrier, according to which acquired traits such as alcoholism cannot be inherited, becomes a problem. This leads him to express skeptical attitudes towards positivist science in his last novel, but this does not invalidate his construct of the experimental novel oriented towards natural scientific laws.

Zola thus adopts, mediated through Claude Bernard, the method of the English empiricists to proceed from experiences and experiments. When he refers to himself as an observer and experimenter in the human realm, he sees humans as physically determined like physical bodies and without free will. If the inductive method and observation are also to be applied to human action, then there is a paradigm shift from the subject to the object, from teleology to causality, from deduction to induction. Ethics, politics, or theology are no longer at the center of interest, but scientific findings that enable future predictions and serve to control nature. The path leads from experimental physics to the experimental novel, which is the opposite of idealistic or romantic novels. However, the experiment presupposes a hypothesis that it confirms or refutes. Zola's hypotheses arise from the laws of inheritance, the environment, and can be derived from the historical moment, as already suggested by the subtitle of his history of the Rougon-Macquart family, *Histoire naturelle et sociale d'une famille sous le Second Empire.*

6.7   MEDICAL VISIONS

Santiago Ramón y Cajal (1852–1934) was the one who, in Spain in a similar way as Emile Zola in France, albeit less systematically, incorporated scientific thinking into literature. He was a physician specialized in microscopic anatomy. For his work on the fine structures of nerves in the brain and spinal cord, he received the Nobel Prize for Medicine in 1906. He also worked as a painter and writer. He wrote his 1905 published *Cuentos de vacaciones* during the cholera epidemic of 1885–1886 in Valencia, which may have contributed to the fact that, unlike with Jules Verne, travels and discoveries are not the main theme, but rather the human body. The subtitle of the collection

---

[110] Klinkert 2020, 252–274.
[111] Föcking 2002, 342.

Narraciones seudocientíficas suggests that the stories are based on facts and hypotheses of contemporary biology and psychology, which is why basic scientific knowledge is required from the reader.[112]

In the revised and expanded publication of his inaugural address, which he gave in 1897 before the *Academia de Ciencias Exactas, Físicas y Naturales*, he mentions observation, experiment, and induction as the essential sources of human knowledge right at the beginning. He rejects aprioristic principles, intuition, inspiration, and dogmatism. Therefore, the knowledge of Descartes, Fichte, Krause, and Hegel, who want to find the laws of the universe and the solution to the great mysteries, such as the beginning of life or consciousness, by exploring their own minds, are chimeras and a waste of time. Rather, one should inform oneself with Galilei, Kepler, Newton, Lavoisier, Geoffroy Saint-Hilaire, Faraday, Ampère, Claude Bernard, Pasteur, Virchow, and Liebig. As these show, the observation of phenomena, description, comparison, and classification using analogies and differences are meaningful in order to arrive inductively at empirical laws. Referring to Claude Bernard, he advises the scientist to limit himself to causalities and not to search for first causes or substances behind the phenomena. In observing changes, the how and not the why is important. For observation to be free of prejudice, it is useful to forget one's own accumulated book knowledge. General sentences should not be used as world formulas, but only as working hypotheses. As such, they can only arise through induction. If they do not stand up in verification or falsification, they are to be discarded. Scientific laboratories and factories may work closely together. In natural science, it is important to work patiently, long, and intensively. Only in this way could Pasteur invent his vaccines. Doomed to failure, on the other hand, are dilettantes, meditators, scholars, book lovers, and theorists.[113]

Ramón y Cajal leaves behind theological and philosophical traditions and places his characters in a context shaped by scientific thinking.[114] His *Cuentos de vacaciones* have an optimistic tone.[115] In the short story *A secreto agravio, secreta venganza*, he adopts the title and the core of the plot of a play by Calderón de la Barca from the 17th century. The protagonist, whom the author, who appreciates the Germany of his time, gives the name Max v. Forschung, is a famous bacteriologist who, however, demonstrates the selfish and amoral side of a scientist.[116] When his significantly younger wife cheats on him with his younger colleague Mosser, he administers tuberculosis bacteria to Mosser, causing him to die. The fact that the wife also becomes infected serves as proof of her infidelity. After v. Forschung has saved his wife

---

[112] Ramón y Cajal 1961, 683.

[113] Ramón y Cajal 1971, 13–17, 31, 79, 84, 116, 120, 126.

[114] Pratt 1992, 1001.

[115] Pratt 1992, 82.

[116] Ramón y Cajal 1961, 684.

with a newly invented serum and the two are reconciled, he gives her a drug that makes her age quickly, bringing her closer to him in age. Since the aging process also involves psychological changes, such as a reduced inclination to criminal behavior and the damping of unrestrained desires, he considers using this drug on a larger scale in the interest of the state and society.

In *La casa maldita*, Inés waits for her fiancé, the scientist Julián, who lost his fortune in a shipwreck returning from Mexico. Nevertheless, he buys a dilapidated house, which he beautifully renovates despite numerous challenges, so that they can marry and live a happy life there. Eventually, he convinces the conservative Inés of the benefits of scientific progress. The plot allegorically shows the overcoming of technically and scientifically backward Spain by the progress-oriented thinking that Julián brought from the New World. The author expressly manifests his sympathy with Julián in the preface.[117]

El hombre natural y el hombre artificial also contrasts traditional thinking with empirical-scientific thinking. While the *hombre natural* has built factories in France as an engineer, the *hombre artificial* deals with metaphysical problems. The former first deals with things and then with books, the latter gets entangled in errors due to lack of experience, since religion, although necessary in the development of human society, must give way to the advancing evolution of natural science.[118] Thus, Cajal aligns himself with the three-stage law of Auguste Comte. One could even describe the *hombre natural* against the background of Freudian theories as a person whose superego is not burdened by unnecessary education, especially of a metaphysical nature.[119]

In the story *La vida en el año 6000*, translated from the unpublished manuscript into English, advances in medicine and human evolution result in the disappearance of religion and philosophy. In the spirit of social Darwinism, human species have evolved that pursue professions corresponding to their organ development. Thus, lawyers have atrophied arms and legs, but still functioning thumbs. Doctors have microscopically sharp eyes and hands as sensitive as those of the blind. Soldiers appear as extensions of their weapons. Medical education, thanks to numerous machines, leads to professions such as the biological precision mechanic or the technical biologist. University teaching, since it repeats itself anyway, is taken over by phonographs. There are no more musicians and composers. Since music has become the photography of natural sounds and the reproduction of phonetic harmony of living bodies, it has become a natural science. Theater has become a collection of photographically captured conversations of gesticulating people, and the novel has become historiography, especially the natural history of mankind.[120]

---

[117] Ramón y Cajal 1961, 685.
[118] Davis 2013, 313–335.
[119] Pratt 1992, 87.
[120] Perez 2017, 204, 211, 223.

In Ramón y Cajal's story *El pesimista corregido,* the scientist Juan falls into a state of general despair after losing his parents to tuberculosis and pneumonia and his fiancée to typhoid fever. As he laments the inadequacy of human perception, a being appears to him, calling itself the "Numen of Science", and grants him microscopically precise vision for a year. After initially rejoicing at being able to recognize and name all microbes, this becomes unpleasant for him as he can clearly see impurities, wrinkles, and makeup on women's skin, so he soon prefers the normal state. Thus, he realizes that he can make more progress in science with measure, patience, and a positive attitude than with supernatural powers. It turns out that not all technological achievements are desirable.[121]

Ramón y Cajal thus shows in his inaugural speech his conviction of the superiority of the natural sciences when he rejects a priori principles, introspection, meditation, inspiration, and dogmatism, and prefers observation, experiment, and induction. One should seek causalities and not first causes or substances, replace book knowledge with patient observation, and consider general statements as hypotheses. Only then can one hope, like the researcher in *A secreto agravio, secreta venganza,* to have the right medication ready for all eventualities. Julian also shows optimism in *La casa maldita,* when he convinces Inés of the progress of natural science. Similarly, the *hombre natural* without books appears superior to the *hombre artificial* entangled in metaphysical problems. Even more advanced in *La vida en el año 6000* are the representatives of individual professions with body parts transformed and optimized in a Darwinian manner, with music, theater, and novels becoming dispensable and disappearing, which Ramón y Cajal does not seem to regret. The job-related transformation of organs in *El pesimista corregido* is not the result of millennia of development, but the fulfillment of a wish by a being called the "Numen of Science". The joy that the scientist Juan, with his microscopically precise vision, recognizes all microbes with the naked eye is clouded by the fact that he also sees unaesthetic things sharply. This is probably less a criticism of scientific progress than a criticism of the impatient scientist. After all, Ramón y Cajal had praised Pasteur's patience as exemplary in his inaugural speech.

---

[121] Comparone 2017, 46–59.

PART IV

CRITICISM OF THE PRIORITY OF THE NATURAL SCIENCES

CHAPTER 7

Life in the Historical Context

The English empiricists admired by Voltaire were the ones who most influentially proclaimed the dominance of the scientific paradigm. One should proceed inductively from experiences and experiments. While in antiquity ethics and politics held dominance, and in the Middle Ages theology, now the attention is on nature, which should then determine ethics and politics through the inductive method. In Adam Smith's impartial spectator, morality became the result of a particular kind of empirical observation. If, building on this, Zola wants to replace the previous metaphysical preoccupation with man, which led to classical and romantic literature, with the consideration of physical and chemical factors of man, and Ramón y Cajal wants to turn music, theater, and novels into natural science, it is not surprising if opposition arises against this.

The criticism of the Spanish essayist and philosopher Ortega y Gasset (1883–1955) of the priority of the natural sciences is systematic. Drawing on early modern humanistic positions, he contrasts scientific thinking with categories such as life, history, imagination, beliefs, past, and change. He sees the development of the natural sciences characterized by increasing specialization and the loss of foundations, which in turn leads to the loss of metaphysics and the impoverishment of human existence. Ortega y Gasset's assessment of the natural sciences must be seen against the background of his personal development. Until 1913 he was a follower of neo-Kantianism, then he approached a phenomenology that subordinates the *razón pura* to a *razón vital*, thus making knowledge and science an instrument of life. From 1929 onwards, history becomes the determining factor for him, with the life of the self seen in interaction with its historical environment, its *circunstancias*.[1] Thus, he

---

[1] Pinto 2009, 198–199.

C. Strosetzki, *Literature in Dialogue with the Natural Sciences*,
https://doi.org/10.1007/978-3-662-71319-8_7

acknowledges the successes of natural science, but points out that nature is only one dimension of human life. The human itself eludes physical-mathematical reason, as it is not a thing and must be thought in categories that radically differ from those that explain the phenomena of matter. Life is a gerund and not a participle, a *faciendum*, not a *factum*. It is the task of a free agent. It is to be brought about. It is not a being-already, but a being-should. The life plan is the self of each individual, which he has chosen from various possibilities. This includes the imagination, the ability to invent one's life image, to constantly rethink the personality one wants to be in the face of different opportunities and circumstances. Human life is therefore not a substance, but change. The respective personal past is a moment of human identity. In the principled unlimitedness of his possibilities, man is a wanderer who only takes his past as baggage. He does not have nature, but he has history, *res gestae*.[2] If for Ortega truth is something historical, then this means that all scientific knowledge arises from the context of historical and cultural circumstances, which form a horizon. He also refers to this approach as holistic, insofar as every historical or human fact is bound to its context, with which it forms a kind of organism.[3] While the book conveys intellectual education and historical traditions, the scientific experiment is related to what is currently materially available. Progressive scientific thinking thus leads to a growing disinterest in tradition, history, and literary history.

Humans always live by and from certain beliefs. Life is an operation that moves in a forward direction, thus it is oriented towards the future. However, the future is not within man's power. All he truly possesses is the past. Ortega wonders to what extent the principle of indetermination which he sees in humans also applies in physics. If the scientist no longer merely observes phenomena and defines them with precise formulas, but by observing the phenomenon also produces it, then observation is production. After all, observation is guided by a respective perspective. Here Ortega refers to Einstein, according to whom the scientist, like the poet or painter, can only create an image of reality, which should be exact, but like Poincaré's conventionalism, is nothing more than a construct. After all, space and time are not absolute, but relative to the perspective of the observer.[4] Doubts are also warranted regarding logic. For Ortega, Gödel's theorem of formally neither provable nor refutable statements means that there is actually no logic and what was considered as such starting from Aristotle was a utopia. And since Russel, Whitehead, and Hilbert we know that logical principles have no reality and thus the basic principles of civilization have become obsolete, which

---

[2] Ortega y Gasset 1964, 23–32; on the importance of history for science: Pinto 2009, 195–223.

[3] Miquel 1992, 135–136, 147.

[4] Pinto 2009, 218–220; Harada 2006, 3–13.

is why new ones need to be invented.[5] While from the 16th to the 19th century, physics working mathematically was the basis of all knowledge and was imitated by other disciplines, this position was shaken at the beginning of the 20th century by quantum mechanics and the theory of relativity. For Galilei, Descartes, and Newton, Euclidean geometry was the basis, while Riemann introduced a new geometry for the macrocosm of large distances, which does not start from the rectangle, but from curves.[6]

Ortega defines technology as the changes that man makes to nature in the interest of satisfying his needs.[7] This is evident when he changes his natural conditions, *circunstancias*, by creating dwellings or means of locomotion. Herein he differs from the animal, which must adapt to its environment. Technology is exactly the opposite, namely the adaptation of circumstances to the human subject. In this context, nature or world are circumstances that are interpreted in terms of their difficulties and possibilities. Modern technology begins with Galilei, Descartes, and Huygens, who interpreted the universe mechanically as a machine, thus displacing notions of ruling spiritual powers. With Ortega, three stages of technological development can be distinguished. The technology of chance characterizes the beginning of human history, where the new is still understood as a dimension of nature and is not consciously created. From antiquity to the Middle Ages, a craftsman's understanding of technology prevailed, inventing instruments for man, who remains the main actor. Only with the advent of the machine, which works independently, does man serve the machine by operating it. He has thus established a realm between himself and nature that he has created, which Ortega calls *sobrenaturaleza*. Ortega contrasts the unlimited possibilities of current technology, which is focused on matter, with human emotional life, which does not only deal with matter: "Pero la vida humana no es sólo lucha con la materia, sino también lucha del hombre con su alma."[8]

Ortega views the scientists of his time, who are focused on specialized fields, critically. Experimental science began at the end of the 16th century with Galilei, continued with Newton at the end of the 17th century, until it could unfold in the middle of the 18th century. The later developments led away from the earlier syntheses towards more and more specialization. With Galilei, it is not the experiment itself that is most important, but rather the aprioristic assumption of laws, which are confirmed or refuted a posteriori by experiments.[9] A history of physical and biological sciences would show how scientists from one generation to the next increasingly restrict themselves, commit to a narrower field of activity, and thereby lose sight of the other

---

[5] Ortega y Gasset 1965, 663.

[6] Miquel 1992, 133.

[7] Ortega y Gasset 1939, 14.

[8] Ortega y Gasset 1939, 100.

[9] Pinto 2009, 207.

parts of science, culture, and civilization. The former interest in the entirety of knowledge is disqualified as dilettantism.

When science is divided into small parcels, anyone can engage with it. There is no need for notions of the meaning or the foundations. Thus, Ortega can compare the average scholar in his laboratory to a bee in the comb of the beehive or a horse going around in circles, serving the drive of machines with its rotations. Whereas people could previously be divided into the knowledgeable and the ignorant, the specialist emerges as a new type, whom Ortega refers to as the learned ignorant, as he knows everything in his field and nothing beyond it. The learned ignorant is reminiscent of the previously mentioned *pedant* of the French 17th century. Thus, he appears in everything that does not belong to his field as ignorant as the broad mass, which is particularly evident when he wants to make himself heard on general questions. Therefore, the number of experts in Ortega's time is greater than ever; the number of educated people less than, for example, in 1850. It is time to promote encyclopedic thinking and insight into historical conditions again. Because civilization is not simply present, like the earth's crust or the primeval forest.[10] If there is already a lack of universally educated natural scientists, then there is an even greater lack of an understanding and insightful public.[11]

For Ortega, the positivist limitation to empirical knowledge leads to the loss of metaphysical insights and thus to the impoverishment of human existence. When the natural sciences encounter an insoluble problem, they set it aside. Philosophy, on the other hand, admits from the outset the possibility that there are insoluble problems in the world.[12]

Ortega sees effects of Darwinism in politics, logic, morality, aesthetics, and in religion. While Darwin proves the effects of changed environmental conditions on humans, it is important to Ortega y Gasset to show how changes in humans change the environment. For example, if the eye is added as an organ, it creates the visible world. Once the eye exists, it creates the laws of optics, while no eye arises from the physical laws. Changes in humans also do not appear to be a matter of centuries, but happen in the shortest times from an *exuberancia vital*. The fact that man is a deficient being and wants to supplement his unsatisfactory equipment is another factor of his changes. Scientific truth is therefore exact for Ortega, but neither definitive nor complete, as a myth would have us believe. Seen in this way, Darwinism is a myth. For Ortega, the entire empirical natural science of the 19th century is a myth, and even an imperialist myth, as it claims to reshape other disciplines in its image.[13]

---

[10] Ortega y Gasset 1957, 163–171.

[11] Fisac 2010, 38.

[12] Pratt 1992, 155–157.

[13] Pratt 1992, 163–169.

Einstein has a special significance for Ortega. If he claims that physical reality does not correspond to Euclidean geometry, this only proves for Ortega that mathematics is only a theoretical instrument that does not grasp the structures of the world. What is relativity for Einstein becomes perspectivism for Ortega, which starts from the individual human being. He therefore rejects pure rationality and proceeds from a vital and historical reason. Subjectivity is also the origin of culture.[14]

The development of the natural sciences is thus characterized by increasing specialization for Ortega, leading to the impoverishment of human existence. The human itself eludes physical-mathematical reason. Life is the task of an actor who wants to create something and himself. The life plan is the self of each individual, which he has chosen as his being-should. Here Ortega asks to what extent the principle of indetermination, which he sees in humans, does not also apply in physics.

For Ortega, natural science is an imperialist myth, as it claims to reshape other disciplines in its image. If Darwinism, which shows the effects of changed environmental conditions on humans, has consequences for politics, morality, and religion, then it is important to Ortega y Gasset to show how the environment changes through changes in humans. Ortega thus rehabilitates concepts such as life, history, imagination, past, and change.

[14] Pratt 1992, 171–172.

CHAPTER 8

Fantastic Literature as Metaphysics

The language criticism from Bacon to Locke to Carnap and Wittgenstein denied the possibility of meaningful sentences in both metaphysics and literature. The Vienna Circle of the 20th century is even credited by Sábato with having seen in metaphysics a branch of fantastic literature, a thesis that Borges made the focal point of his literary work: "El Círculo de Viena sostuvo que la metafísica es una rama de la literatura fantástica. Y ese aforismo que enfureció a los filósofos se convirtió en la plataforma literaria de Borges."[1]

Indeed, the Argentinian Jorge Luis Borges (1899–1986) incorporates elements of metaphysics into his stories. Since he knew the work of Ramón y Cajal, he could have been inspired by his *El pesimista corregido* when in his story *Funés el memorioso* he gives his protagonist, as a result of an accident, a comprehensive perception and an infallible memory. Although these gifts have overcome human weaknesses and deficiencies, the price for grasping all details directly is the loss of the ability to think. Because thinking means generalizing. And abstracting means disregarding and forgetting details. Limited to the absorption of new details again and again, he succumbs to insomnia and loses any possibility of creative thinking.[2]

Thinking also means classifying and defining by *genus proximum* and *differentia specifica*. When Borges quotes from a "certain Chinese encyclopedia", one gets the impression of a satirical confrontation with Linnaeus' attempts to classify the animal world in the 18th century: "Animals can be grouped as follows: a) animals that belong to the Emperor, b) embalmed animals, c) tamed, d) suckling pigs, e) sirens, f) fabulous animals, g) stray dogs, h) those belonging to this group, i) those that behave like mad, j)

---

[1] Sábato 1971, 244.

[2] Novillo-Corvalán 2015, 23–44.

© The Author(s), under exclusive license to Springer-Verlag GmbH, DE, part of Springer Nature 2025
C. Strosetzki, *Literature in Dialogue with the Natural Sciences*,
https://doi.org/10.1007/978-3-662-71319-8_8

innumerable, k) those drawn with a very fine camel hair brush, l) and so on, m) those that have broken the water jug, n) those that look like flies from a distance."[3] Foucault quotes this in the preface to his book *The Order of Things* as an example of the attempt to bring empirical details into order.[4]

While here the fantastic lies in the play with subjective thinking, in other stories objective institutions and circumstances are varied in such a way that they do not seem far removed from science fiction. In Borges' *La lotería de Babilonia*, one can not only draw winning tickets in the lottery, but also those that oblige the paying of fines. Players who fail to pay the fines are sent to prison, with the lottery company becoming increasingly powerful. With its mechanisms, it appears as a state within a state and as a parable of a totalitarian state.

When Borges' stories deal with mathematics, logic, and physics, he is concerned with the fundamental problems that have been the subject of philosophy, especially metaphysics, since antiquity. In the story *El Aleph*, the first letter of the Hebrew alphabet becomes the point at which the entire universe is concentrated. It is the point that contains all others, a symbol for God and infinity, and in the mystical Jewish tradition of Kabbalah, a key to understanding the world. Elements of the controversy between nominalism and realism can be found in *El Zahir* and *El Aleph*. In medieval scholasticism, the question was asked whether general concepts have an independent existence separate from individual realities and thus precede them (*ante res*), or whether they only exist in connection with individual beings (*in rebus*), or whether they are merely names for summarizing similar things and are only subsequently (*post res*) formed by us through abstraction. At that time, the nominalist direction had turned against Platonic concept realism and did not regard general concepts as related to real entities, but as names of things.

In another case, a mathematical problem becomes a metaphysical one. In *Los avatares de la tortuga*, the thematization of the relationship between infinity and finiteness ties in with the paradox of the pre-Socratic Zeno, who had claimed that a fast runner like Achilles could never catch up with a tortoise in a race, because whenever he reached its position, it had already moved on again. Where Borges deals with infinity, there are borrowings from Georg Cantor's set theory and the philosophical discussions on the foundations of mathematics by Kurt Gödel.

In *La biblioteca de Babel*, the library is portrayed as eternally existing and the number of orthographic characters is set at 25. Borges refers to the Hebrew alphabet and adds a period, comma, and space. He then develops Kabbalistic notions of creation as an infinite chain of variations of combinations of this finite number of basic elements, asks about the librarian, the book of books, and recognizes in the cycle theory and in the labyrinth the appropriate symbolism for book and library. The question also arises about the relationship between space and time, with time, as also by Kant, being seen as a mental construction of the cognizing subject. As evidence that reality depends on the subject, the insight of the theory of relativity is used, according to which the observation of a tiny object is not possible without it being changed by the observation through the measuring instruments. The resulting dominance of the subject leads to an "idealist", i.e., constructivist perspective in *Tlön, Uqbar, Orbis Tertius*, which is linked with the language criticism of Fritz Mauthner. Language cannot depict reality, but is only a mirror of mental processes.[5] The metaphysicians of Tlön do not seek truth, not even probability, but only the astonishing. Thus, in *Tlön, Uqbar, Orbis Tertius*, a description is made based on fictional sources, making Tlön appear as a place where Borges' views are synthesized. The prevalent belief is philosophical idealism. Since the mind is portrayed as a great whole, the names of authors do not need to be mentioned in literature. Because literature is at the same time a guided dream, a *sueño dirigido*, it exists as a totality independent of historical reality. The world becomes a sequence of mental representations, a constant flow of arising and passing away. Literature thus exists as an objective totality independent of the writing subject, within which the individual author only changes commas or minor things.

Where Borges thus draws on mathematics or physics, it is a matter of fundamental philosophical questions that traditionally overlap into the realm of metaphysics. Thus, his relationship to the natural sciences does not appear comparable to the admiring attitude of Balzac or Zola, but is rather critical and interested in fundamental questions that lie at the boundaries of philosophy and metaphysics.

The coordinates of time and space become confused in Borges' narrative *Pierre Menard, autor del Quijote* just as fiction and reality do. The French protagonist Pierre Menard, who wrote various texts at the beginning of the 20th century, has set his mind on reinventing Cervantes' novel *Don Quixote* without changing the original template. As preparation, he has to learn the Spanish language, familiarize himself with Catholicism, fight against Moors and Turks, and forget European history from 1602 to 1918. An example is given of a passage that shows Menard's text to be identical to that of Cervantes. Yet it is said that Menard's text, despite its external identity, is infinitely more complex and richer. Thus, the external reality seems to mislead, and only the subjective interpretation recognizes the differences between the two texts when reality, time, and texts are constructed by the subject. Borges hints at this in *Otras inquisiciones* when he writes in verses: "Time is a river that sweeps me away, but I am the river, it is a tiger that tears

---

[5] Merrell 1991, 144, 236; see also Martinez 2003.

me apart, but I am the tiger; it is a fire that consumes me, but I am the fire. The world, unfortunately, is real; I, unfortunately, am Borges."[6]

The Argentinian Adolfo Bioy Casares (1914–1999) was friends with Borges. The coordinates of time and space are also central to his work in *La invención de Morel* (1940), although here a technical invention plays a central role. This fantastic narrative is written like a diary in the first person. The protagonist flees to an unknown island. The island seems to be afflicted with a disease, as the vegetation grows faster than elsewhere. He finds a museum with a library, a long chapel, and a swimming pool. On one of his excursions, he encounters people and sees a woman. She meets with a tennis player and is named Faustine, as the protagonist finds out. He speaks to her. But she does not answer. It is as if she has ears and cannot hear him, and eyes and cannot see him. He falls in love with her. The tides, the protagonist believes, seem to provide energy to the engines and generate electricity. Whenever there is light, the human figures appear. He observes Faustine at a meal and notices that they are talking about immortality. He attributes the phenomenon of seeing two moons or two suns on the island at times to a reflection. At first, he considers himself abnormal, then he believes he is invisible, then again he thinks the others are beings of a different nature, e.g., from another planet. Finally, he learns from the inventor Morel that he has filmed the figures and that they now live in the film like in a cinema. They move in a scenario that spans seven days, the time in which everything from sounds, taste, smells to temperature is archived and synchronized on the film. The device can perceive, record, and reproduce in projection. The figures appear through the apparatus and disappear when it is switched off. Since the images have the ability to animate their figures, it seems to be a completely new kind of photography. Morel had bought the island to conduct experiments with electricity. When the narrator now realizes that the beings are projections of an apparatus, he decides to become part of these artificial projections himself in order to be with Faustine.

The question arises whether *La invención de Morel* belongs to fantastic literature or to the genre of science fiction if one defines science fiction as fictional narratives in which scientific and technical inventions play a central role. Narratives reminiscent of later science fiction stories have been handed down from antiquity. For example, in the 2nd century AD, Lucian of Samosata described space travel in his *True Stories*. On the moon, the travelers learn that the king of the moon is waging a war against the king of the sun with an army of mushrooms and centaurs, the dispute being over the morning star. Back on earth, where people are fighting each other, they are swallowed by a whale. Finally, they meet the figures of the Trojan War on an

---

[6] Borges 1989–1996, 146.

island of the blessed. With these stories, Lucian wanted to satirically criticize historians who present myths as historical truths.

In the Middle Ages, the philosopher Roger Bacon anticipated technical developments that later became subjects of science and science fiction literature:

> "Machines will be built with which the largest ships, controlled by a single person, will travel faster than if they were crammed with rowers; carriages will be built that will move at incredible speed without the aid of draught animals; flying machines will be built with which a man will dominate the air like a bird; machines will allow us to reach the bottom of seas and rivers."[7]

When scientific and technical inventions play a central role in science fiction stories, four groups can be distinguished. First are travel stories in the tradition of Lucian. The most famous are the travels in time and space of the early science fiction author Jules Verne (1828–1905). In space, the traveler can also encounter aliens. Here, reality is shaped in such a way that humans are confronted with themselves as objects. Secondly, the subject of man himself can be changed, as in Cajal through medicine. The changes can lead to cloned or immortal humans or to humans who can read minds. Thirdly, technology can take center stage when robots or artificial intelligence come into play, dominating and disempowering humans and making decisions for them. Such a society then appears as a surveillance state or as a dystopia. Fourthly, a post-apocalyptic state of origin without technology as a result of a nuclear war or a natural disaster can become the subject.

So if in antiquity travels to the moon and in the Middle Ages flying machines still seemed fantastic, they become reality with the advances of natural science over the centuries, so that science fiction stories present new scientific and technical inventions. They present a new reality, the changed man himself, man facing a dominating technology, or man in a post-apocalyptic state of origin. Borges' fantastic stories thematize the subject, whose perception and memory are perfect, but whose thinking and generalizing are lost. How complicated classifying and defining is, Borges shows with his Chinese encyclopedia. Do general concepts have a reality independent of the respective realities? This question is dealt with in *El Zahir* and *El Aleph*. And why can't Achilles overtake the tortoise? The eternally existing Library of Babel raises questions about creation as an infinite chain of variations of combinations of a finite number of basic elements. Space and time appear in Borges not only like in Kant as a mental construction of the recognizing subject. If language is considered a mirror of mental processes, then even a philosophical idealism is represented. An example of an idealism in which the coordinates of time and space as well as fiction and reality are abolished is Borges' *Pierre Menard, autor del Quijote*.

---

[7] Bacon 1618, 37.

The question arises whether Bioy Casares' *La invención de Morel* belongs to fantastic literature or to the genre of science fiction, if one defines science fiction as fictional narratives in which scientific and technical inventions play a central role. After all, the characters are tied to a device, a "soulful" photograph, and disappear when it is not running. But if the protagonist himself wants to become part of these artificial projections in order to be with Faustine, here too the boundaries of time and space as well as fiction and reality are crossed.

CHAPTER 9

Language and Natural Science

If literature and art, like metaphysics, are expressions of lived perception and not of knowledge, and do not produce meaningful or verifiable sentences, then two possible reactions arise. One can, like Borges, base these views on a poetics in which philosophical metaphysics becomes the foundation of literary shaping and experiencing, or one can, like Sábato, refute them.

For the Argentine writer and scientist Ernesto Sábato (1911–2011), literature is about knowledge, especially in metaphysical questions.[1] In contrast, mathematical symbols appear to him as a marble museum. The individual asks what the scientific effort to control the world is for if the mysteries of life remain unsolved. Essential, on the other hand, are metaphysical questions about the meaning of life, the soul, or God: "¿Tiene algún sentido la vida? ¿Qué significa la muerte? ¿Somos un alma eterna o meramente un conglomerado de moléculas de sal y tierra? ¿Hay Dios o no?"[2] In contrast, the law of gravity, the steam engine, and the Kantian categories are child's play. If transience is a central problem, then this is emphasized nowhere as much as in Latin America.[3]

The questions of the individual cannot be answered by logic and natural science precisely because they operate in the general and not in the individual concrete. In contrast, the poet already possesses an individualizing tool with his linguistic style.[4] The language of literature is capable of being metaphysical and conveying lived experiences. Each of the great novelists like Balzac, Dostoevsky, or Proust conveys a worldview and thoughts about the world

---

[1] Campa 1983, 17.

[2] Sábato 1988, 61.

[3] Sábato 1971, 39.

[4] Sábato 1971, 207.

© The Author(s), under exclusive license to Springer-Verlag GmbH, DE, part   199
of Springer Nature 2025
C. Strosetzki, *Literature in Dialogue with the Natural Sciences*,
https://doi.org/10.1007/978-3-662-71319-8_9

and human existence. While the scientist in his treatises only conveys conceptual skeletons of reality, the writer offers a complete picture, provides insights into existential matters, and shows the truth as he perceives and experiences it. "Nos ofrece una significación."[5] Thus, the poetic approach to truth is not through proof or propaganda, but through a meaningful lived experience, as Carnap would also say. Sábato asks whether poetry, like natural science, needs a specific new language. However, he considers a separation of everyday language and poetic language to be wrong, as there are no poetic words, but only poetic facts, which should be expressed with as simple and transparent a language as possible, i.e., with everyday language.[6] The simplicity is precisely the result of special effort.[7] The *novela metafísica*[8] Sábato considers to be the appropriate genre. His definition of the *dramas metafísicos* again makes it clear that, like Carnap, he conceives of metaphysics and literature primarily as lived experience. His narratives are about the human condition, how he experiences loneliness, death, hope or hopelessness, striving for power, the search for the absolute, the meaning of life, and existence or absence of God.[9] Literature has in common with metaphysics and religion the fact that it primarily deals with fundamental questions, which is why criticism and rejection of metaphysics is often associated with the rejection of literature.

The reality essential for literature is thus for Sábato not an objective one, but a subjectively experienced one and as such linguistically recorded.[10] There is no objective reality without subjective imprint, not even in art. Sábato criticizes the poetological ideas of the realists, who wrongly assumed that there is a reality outside of man that can be known, described, or drawn independently of subjective perceptions and insights. With regard to Stendhal's realism, he asks what the exact reproduction of the outer world is good for, if it is possible at all. Whoever tries this overlooks that man is a kind of tracing paper that gives his own color to reality.[11] At the beginning and at the end stands the subject, as Sábato clarifies in various aphorisms. Whether someone is traveling in foreign lands or exploring nature, in the end he realizes that what he was looking for was himself.[12]

---

[5] Sábato 1971, 262.

[6] Sábato 1971, 207.

[7] Sábato 1971, 209.

[8] In Sábato's metaphysically influenced artwork, Marianne Kuener recognizes approaches of existentialism, but also the aesthetic and philosophical dimension that had connected Romanticism to the total work of art: Kuener 1991, 252–253.

[9] Fernández 1983, 35.

[10] Fernández 1983, 38.

[11] Cersosimo 1992, 194; Kasner 1992, 112.

[12] Sábato 1981, 15, 58.

Thus, the literature of an era does not reflect objective facts, but subjective "worldviews" of its contemporaries.[13] Such a subjectively influenced conception of art then fits Sábato's recourse to Hegel, with which he emphasizes the special rank of art, whose manifestations have more reality and truth than the phenomena of the real world.

The preference for the subjective over the objective in life and in the conception of literature also corresponds to the priority of experience over reflection[14] and the idea that the novel conveys knowledge as lived experience in a realm between ideas and passions.[15] Thus, according to Sábato, metaphysical fears cannot be imagined as pure ideas, but only dressed in feelings and passions.[16] From philosophical, it becomes a psychological metaphysics, from a general treatise, a concrete novel.

As much as Sábato emphasizes the advantages of metaphysical literature, he criticizes the shortcomings of natural science. Its method of precise observation and logical analysis certainly has validity. However, if one considers the findings, one must realize that they are only of temporally limited validity. For example, the teachings of Ptolemy were overcome by Copernicus. Einstein brought a new correction, which would also be corrected by a more complex theory. The progress of natural science can always be traced back to such dialectical negations. Even though the layman admires the power of natural science and venerates its representatives like Albert Einstein or Marie Curie, experts express doubts that the increasingly growing abstraction of natural science brings with it growing power. This worries the experts so much that they begin to doubt whether natural science is still capable of grasping reality. The mathematical-logical language postulated by the Vienna Circle runs the risk of becoming autonomous and producing another kind of fantastic literature. According to Sábato, natural science came to its power through a pact with the devil, which led to the evaporation of the everyday world. Thus, natural science has become the sole ruler, but over a realm of ghosts.[17]

Not only because of its abstractness does the language of natural science suffer a loss of reality. Sábato quotes Russel, for whom physics is not equated with mathematics because we know a lot about the outside world, but because what we know about it is very little. But how could necessity and exactness as properties of logic and mathematics succeed in psychology, where they are hardly suitable for grasping physical reality, Sábato asks. Precisely because of their mathematical form, valuable areas of human reality are lost to natural science. Feelings or the sense of justice can no more be put

---

[13] Sábato 1988, 67.

[14] Sábato 1971, 200.

[15] Sábato 1971, 203.

[16] Sábato 1971, 14.

[17] Sábato 1981, 27–28; see also 44, 26.

into mathematical form than the ideas of a beautiful palace, a beautiful land-scape, or the music of a fugue by Bach.[18]

The idea of the Vienna Circle that the mathematized language of the nat-ural sciences enables progress, while the languages of metaphysics and liter-ature have not visibly advanced for centuries, is taken up by Sábato, albeit not without relativizing progress and interpreting the supposed standstill as a diversity of equally valid designs. If every epoch and every country finds the language that best expresses its *pathos* and *ethos*, then only in the fight against the generalizations of conceptual language. The language of natural science, such as the Pythagorean theorem, which establishes a relationship between the three sides a, b, and c in a right-angled triangle, remains for-ever alien to the values of ethics and aesthetics with its formal propositions. Sábato agrees with the Vienna Circle when he regrets that it was detrimen-tal for natural science to use the words of everyday language to symbolize abstract facts. Since these words had connotations from the world of life and were emotionally charged, they had to hinder the progress of thought. Thus, for Sábato, it was a justified consequence that emotional contamination was ended in the natural sciences by agreeing on a certain number of symbols that had no other meaning than the one their inventors had given them at inter-national scientific congresses.[19] Such a logical artificial language is deliberately distinguished from everyday language, which is as illogical as the everyday world, which should not only formulate abstract knowledge but also express feelings, influence others, or instill sympathy or antipathy in them. With its contradictions, allusions, and absurdities, everyday language is the basis of the language of literature.

This is exemplified in the significance of the metaphor. The very fact that it means something other than it signifies is its advantage. It is not mere deco-ration or rhetorical accessory, but a good way to express the subjective world. The metaphor is a prime example of the expression of subjective experience. It is particularly true for it that it is unsuitable for conveying the truths of logic and mathematics, but it is the appropriate linguistic means for the truths of subjective existence with its convictions, hopes, and fears. Although the metaphor does not depict reality, but equates dissimilar things, it has not only a psychological but even an ontological value for Sábato, as it is capable of illuminating the deepest layers of reality.[20] With its subjectivity and concrete-ness, it represents the extreme contrast to the mathematical-logical language of the natural sciences, which is general and objective, but alien to the world and unsuitable for metaphysics and literature alike. Thus, Sábato shares with the Vienna Circle the conviction of the necessity of a logical-mathematical

---

[18] Sábato 1981, 30–31.

[19] Sábato 1971, 202.

[20] "Tiene un valor ontológico, que actúa por alumbramiento de los estratos más profundos de la realidad." Sábato 1981, 100.

language for natural science. He also adopts the idea of the similarity of literature and metaphysics, but not to make metaphysics a literary game like Borges, but to expect the really important insights precisely from the combination of literature and metaphysics. With the same premises, the Vienna Circle thus results in an upgrading of natural science and a downgrading of literature and metaphysics, while Sábato devalues the natural sciences and upgrades literature and metaphysics.

Borges, on the other hand, takes a different path according to Sábato, when he deals with philosophy in a playful and eclectic way. Out of aesthetic interest, Sábato believes, he seeks the peculiar, the entertaining, and the astonishing, such as logical paradoxes, *regressus ad infinitum*, which he uses for his stories. Eclectically, Borges can draw on Parmenides or George Berkeley. This is unproblematic for him, as he is not concerned with finding truth.[21] He goes through the world of philosophical thinking like a collector through an antique shop. Intellectual games with invented worlds without reference to reality fascinate him. He does not seek philosophical or metaphysical knowledge. His rhetoric makes him a sophist, not a philosopher, according to Sábato.

He does not want to participate in the ever-difficult process of truth-seeking. Like a sophist, he uses evidence and debates for the sake of debating. He takes pleasure in this. As a writer, he most enjoys creating words upon words like a sophist: "La discusión con palabras sobre palabras."[22] What Borges writes in stories like *Tlön, Uqbar, Orbis Tertius*, neither he nor his readers can believe to be true, although the metaphysical implications are pleasing: "Aunque a todos nos encanta lo que tiene de posibilidad metafísica."[23] And so it is when he speaks of the world as a repeatable dream, of immortality through transmigration or through the memory of others, or assures that these only exist in eternity. Everything is equally valid, although strictly speaking nothing is valid.[24] Thus, metaphysics and fantastic literature become equivalent. While the Vienna Circle equated the epistemic value of metaphysics with that of fantastic literature and derived consequences for a reform of philosophy in favor of future knowledge gain, Borges did not care about remedies, turned necessity into a virtue, and created fantastic literature with philosophical elements.

Borges was influenced not least by the philosopher and language critic Fritz Mauthner (1849–1923), who was convinced that language would be the most important future topic of philosophy. He opposes both the idealistic speculations of Hegel and materialism or positivism of Comte. Like Wittgenstein, he practices reason criticism as language criticism, even though

---

[21] "ya que él no se propone la verdad." Sábato 1971, 245.

[22] Sábato 1971, 246.

[23] Sábato 1971, 246.

[24] "Todo es igualmente válido y nada en rigor vale." Sábato 1971, 247.

Wittgenstein's *Tractatus* is closer to Russel than to Mauthner.[25] Borges himself is said to have stated that Mauthner's *Dictionary of Philosophy* (1910) was one of the works he repeatedly read and annotated. Indeed, there are three agreements between Borges and Mauthner in their conception of language: Both see language as an arbitrary symbol system, emphasize the social character of language, and deny the possibility of language as a representation of reality.[26] For example, where Borges cites encyclopedias in *Tlön, Uqbar, Orbis Tertius*, he aims to criticize the idea that they provide a complete image of the real world.[27] In this, he follows Mauthner, who states in view of the temporality of every knowledge system: And so I came to the conviction: there can be no objective system of knowledge, even the utmost reflection must remain subjective human work.[28] Mauthner formulates more generally: "Wissen ist Wortwissen. Wir haben nur Worte, wir wissen nichts."[29] Although it is possible to compare ideas and concepts with each other, it is never possible to compare them with the thing in itself. For Mauthner, knowledge is the longing to get beyond language, or the illusion of metaphysicians to have gotten beyond language.[30] The question of whether it is possible to transcend the boundaries of language, literature, consciousness, and imagination to reach the world in itself seems to be what Borges is asking.[31]

Borges leaves the answer to this question open, deliberately blurring the boundaries between the object level and the meta-level, reality and fiction, and between literature and philosophical metaphysics. He seems to move in a labyrinth where literature, metaphysics, and poetic creation are not separated from each other, but are paths in search of an exit that does not exist.[32] In this respect, Borges draws different conclusions from the critique of traditional metaphysics than logical positivism. While the latter tried to create a new, logically exact language, Borges retains the inadequacies of metaphysical language and emphasizes their fictional character by making them elements of narratives. He is to be understood in the same way when he considers metaphysics as a branch of fantastic literature: "La metafísica es una rama de la literatura fantástica."[33]

---

[25] "Alle Philosophie ist 'Sprachkritik' (Allerdings nicht im Sinne Mauthners.) Russels Verdienst ist es, gezeigt zu haben, daß die scheinbare logische Form eines Satzes nicht seine wirkliche sein muß." Wittgenstein 1969, 26.

[26] Echevarría 1983; Dapía 1993, 29.

[27] It is certainly simplifying to label Borges' language criticism as nominalistic, as the nominalist sees general concepts as conventions, but does not deny the possibility of describing or explaining the world: Rest 1976, 50.

[28] Mauthner 1910, 396.

[29] Mauthner 1918, 231.

[30] Kühn 1975, 64.

[31] Benavides 1992, 260.

[32] Gutiérrez Girardot 1992, 296.

[33] Borges 1974, 436.

His fantastic worlds arise from grammatical experiments with metaphysical implications when he designs worlds without memory and without time or contemplates the possibility of a language without nouns but instead with impersonal verbs and indeclinable epithets.[34] As a writer, he is influenced by numerous authors before him. He adopts language as a tradition, as a kind of world feeling, and not as an arbitrary repertoire of symbols.[35] His narratives are characterized by a learned game with existing or invented intertextual references.[36] And each individual narrative is again polyvalent and eludes clear assignment.[37] This becomes particularly clear with the metaphor, whose central importance the early Borges of Ultraismo emphasizes: "Hemos sintetizado la poesía en su elemento primordial: la metáfora, a la que concedemos una máxima independencia."[38] It also becomes clear that literature is primarily to be understood as a subjective experience, and not as a statement about reality.

In contrast to the language critics of the Vienna Circle, Sábato believes that literature conveys knowledge, especially in the area of essential questions of metaphysics, while natural science leaves the mysteries of life unsolved. While natural science, according to Sábato, moves in the conceptually general, literature is devoted to the individually concrete and the human condition. Worldviews and experiences ensure that there is no objective reality without subjective imprinting. The mathematical-logical language, on the other hand, leads to the evaporation of the everyday world and produces, in a certain sense, fantastic texts. A sentence like that of Pythagoras remains alien to the values of ethics. For their *novelas metafísicas* and *dramas metafísicos*, literature uses everyday language, which is also the language of poetry. While Locke rejected figurative expressions, for Sábato and Borges it is precisely the metaphors that are capable of illuminating the deepest layers of reality.

Borges, from Sábato's point of view, is not concerned with finding truth when he equips fantastic literature with philosophical elements. When he makes metaphysics a branch of fantastic literature, he seems to adopt Mauthner's thesis that knowledge is word knowledge and we only have words, but know nothing. Indirectly, he thus confirms the assessment of everyday language and literature that language critics from Bacon to Carnap have formulated, and stands in diametrical opposition to Sábato's view.

---

[34] "Pensé en un mundo sin memoria, sin tiempo; consideré la posibilidad de un lenguaje que ignorara los sustantivos, un lenguaje de verbos impersonales o de indeclinables epítetos." Borges 1974, 539.

[35] Borges 1974, 1081.

[36] Blüher 1995, 119–131.

[37] Fleming 1993, 115–118.

[38] Videla 1871, 203.

CHAPTER 10

Résumé and Outlook

It has been shown that in the early modern period, the paradigms of literature were dominant. A vivid example of this is the metaphor of the world as a book. Literature originated from the subject and thematized its thoughts, doubts, goals, and priorities. Based on this, it could derive details. The nature of things is not revealed through observation, but by answering the question of what they actually consist of. This was to be read in Aristotle, who, like other ancient authors, enjoyed a renewed appreciation during the Renaissance and became the leading authority in the School of Salamanca. When Aristotle explores nature, he starts in physics with basic concepts such as motion, cause, or matter, and in metaphysics with principle, essence, time, part, whole, or necessity. In this context, metaphysics thematizes the most general properties of things and not their accidental variable appearances. Concrete conclusions can be drawn from the general through deduction.

Nature was considered the entirety of material objects that arise and pass away, increase or decrease quantitatively, change qualitatively, or move in space. Other changes are the realization of the possible or transformations and property changes of the given. The five elemental substances earth, water, air, fire, and ether form the basic building blocks, which differ from each other through heaviness and lightness, heat and cold, dryness and moisture, and which occur in the objects of nature in different mixtures. Man, who for Aristotle is the measure of all things and the ultimate purpose of nature, is alone in possession of reason, which is why he enjoys special appreciation. When the founder of Baroque scholasticism Francisco Suárez picks this up, distinguishes between substance and accident, causal and final cause, and accidents such as quality, habit, time, and place, and makes these the first principles of cognition, then metaphysics is the prerequisite for physics. Subjective metaphysical speculations based on ancient texts were the basis for the knowledge of nature.

© The Author(s), under exclusive license to Springer-Verlag GmbH, DE, part   207
of Springer Nature 2025
C. Strosetzki, *Literature in Dialogue with the Natural Sciences*,
https://doi.org/10.1007/978-3-662-71319-8_10

On a completely different, but comparable level, are the speculations about the **similarities** of things, which result from psychological projection and have been handed down in numerous texts since antiquity. Here, insights into the nature of things are not derived by deduction from the most general facts or the basic building blocks, but by attributing similarities. The basis is the correspondences between the microcosm of man and the macrocosm of the universe. If the seven planets are assigned the seven metals or the planets are connected with colors, and if the harmony of the spheres of the seven planets corresponds with the seven notes of the scale in music, then these are not observations, but speculations of subjective thinking. This is also the case when Saturn, because it is furthest from the sun and as the planet of lead, standing for darkness and depth compared to the gold of the sun, is associated with melancholy and stands for the knowledge of secret things. While in the School of Salamanca deduction starts from the most general facts of nature, from the basic building blocks like the first causes, alchemy uses the attribution of similarities, from which it draws insights into the nature of things. In both cases, the observation of the individual is not the starting point, but individual things are derived speculatively based on general assumptions.

If one does not consider nature in general, but specifically the **nature of man**, then the first glance in the early modern period does not fall on his physical constitution, but on something incorporeal like his goal. Further characteristics are derived from his goal, happiness. Since happiness can only be realized in a state-organized society, man is a social being. Since happiness is further only achieved through virtue, i.e., the right middle between two extremes, he has to act ethically. He is defined not by his material components, but by his actions and goals. This is not only the view of Aristotle and his early modern Spanish recipients, but also of French humanists like Montaigne. Therefore, Montaigne rejects grammarians and humanists, who are only interested in irrelevant details, and prefers ethical content. Finally, he considers man as an acting subject, which should be led to the goal of the right life through education. Readings should promote self-understanding through understanding others and clarify one's own situation. Understanding books, therefore, plays as central a role as understanding in oral conversation. Understanding is a procedure in dealing with people and their literary constructs which differs from the scientific explanation of causal phenomena of the object world and characterizes man as a social being. For Montaigne, everyday communication is comparable to dealing with books. After all, one learns not only from books but also in conversation when dealing with behavioral norms inevitably leads to the correction of our own behavior. Whatever is told, it is only valuable if it leads one's own motivations and goals on the ethically right path. The nature of man, therefore, reveals itself in Montaigne through its goals, actions, and norms.

Natural science explains the objects of the world through laws. If, as in the early modern period, one views the **world as a book**, the intention behind it is to understand its meaning. The general presence of the book had increased thanks to printing, making metaphorical use promising. After all, the book of the world holds lessons that are also accessible to the illiterate. Just as the material aspects of paper and ink are not important in a book, the book of the world is not about the superficially visible, but about the hidden message behind it. According to Luis de Granada, one can read the author's wisdom from the book of the creatures of the world. For Antonio de Torquemada, the orderly course of the stars, the effects of the sun, moon, and other planets, testify to a clear conception. However, this is not only harmonious, as Luis Alarcón makes clear when he refers to the positive in the world as "books of God" and the negative as "books of the devil". So if the world is viewed as a book, it programmatically indicates that a general understanding of meaning is important, not observing individual phenomena to subject them to scientific experiments.

Natural laws are known as components of physics, such as the law of gravity, with which Isaac Newton explained gravity on earth and the movements of the planets around the sun in his *Philosophiae naturalis principia mathematica* (1687). However, there are also laws in jurisprudence. If laws are hierarchized in antiquity, then the subordinate level is to be derived from the superordinate one. From the *ius divinum*, a *ius naturale* is to be derived, which should be the standard for the laws of positive law. The Stoics, who believed in a divine world reason, distinguished between the eternal world law (*lex aeterna*), the natural law (*lex naturalis*), and the man-made law (*lex humana*).

If in the Middle Ages Thomas Aquinas assumes a common order for creator and creature from a Christian perspective, then the *lex Dei* applies on three levels: in nature, in the laws of the state, and in the moral commandments for the individual. If the individual acts morally well, then he aligns his individual reason with the reason of creation. From the *ius divinum*, a *ius naturale* can be derived, which in turn is decisive for the individual laws of positive law. According to Thomas, *lex humana* should be derived as positive law from the two superordinate levels, *lex aeterna* and *lex naturalis*. The tensions and contradictions resulting from the different levels of abstraction are conscious to Suárez in the *Siglo de Oro* when he distinguishes between three levels of natural reason: the general principles of moral action, the special principles, and the conclusions.

In early modern Spain, the natural law, *lex naturalis*, was assigned to both ethics and jurisprudence. Here too, the guidelines of Aristotle are valid for the School of Salamanca. If man as a *zoon politikon* needs good constitutions that take the middle into account, then the golden mean between too much and too little is a basic principle derived from the proportions of nature, which applies equally to jurisprudence and ethics. By prohibiting actions, the law generally promotes virtue. Courage appears as the middle between recklessness and cowardice, justice as the middle between doing wrong and suffering wrong, while injustice is too much of an advantage and too little of a disadvantage.

Starting from an ordered creation, in which everything strives for the good, for Thomas Aquinas the highest commandment and law is that the good is to be done and striven for, from which all other commandments of natural law are to be derived. Against this background, the Ten Commandments of the Decalogue also appeared so universally valid that they were read as a natural law text, although they were further abstracted and reduced to the commandment of love in the New Testament. Benedict of Nursia anticipated the categorical imperative of Kant when he formulated that one should not do to anyone what one would not want to suffer oneself. Not physics, but ethics and jurisprudence seem to be in the foreground in the texts handed down in early modern Spain when laws are derived from nature. The subject and his action in the community are prioritized.

When Erasmus of Rotterdam advises modesty in knowledge, he argues from the standpoint of ethics. Vives' concept also originates from ethics, expecting the scholar to maintain his stoic calm undisturbed by external influences and passions such as greed for money, and to use his knowledge modestly for his life and for general education. In doing so, on the one hand he must keep himself away from public opinion, but on the other hand, he must fulfill his moral mission to educate the ignorant. Wisdom is associated with **knowledge**, but knowledge in the case of, for example, chiromancy, pyromancy, and judicial astrology is not associated with wisdom. For Vives, Christian wisdom is superior to the secular wisdom of antiquity. Since the wise person has ethical criteria for the relevance of objective knowledge, wisdom originates from the subject and not from the object.

The important **category of understanding** for literature is the one that is central to Montaigne. In the physical world, causes and effects are explained by laws. With understanding, on the other hand, one encounters other people, their actions and thoughts, but also oral expressions and books. When understanding deals with one's own self, it leads to self-understanding. Montaigne notes the diversity of opinions and constant changes, but does not become a skeptic, instead developing a complex theory of understanding, according to which one's prejudices and prior information guide the understanding of others. In this process, foreign habits that contradict one's prejudices can expand one's personal horizon. This applies not only to encounters on distant journeys, but also to obscure fields of knowledge, where one naturally has to beware of deception. Through merging horizons and application to one's own circumstances, the initially foreign is perceived as one's own. The concept of entirety is constantly to be corrected by the individual elements to be experienced, just as the individual elements appear in a new light through the then also modified entirety. This process, which is present both in understanding foreign actions and in understanding books, has later been referred to as the "circularity of understanding". Even when isolated quotes from the whole book make a circular approach difficult, Montaigne appreciates quotes that he can make his own. He does not think much of aids to understanding such as comments, as they only replace one word with another and may thereby only increase the ambiguities. However, he anticipates the idea that it is grammatical ambiguities that cause most disputes in different scientific fields. Montaigne anticipates categories of 19th and 20th century hermeneutics when he makes understanding the central category in dealing with human actions, views, or ideas.

What knowledge is relevant depends on the criteria by which it is judged. Without such criteria, one is open to everything new, can get to know it with **curiosity**, and strive for encyclopedic knowledge. However, if one makes the familiar the criterion, then the foreign is irrelevant, the philologist despises the natural scientist, the Cartesian despises the specialists, and the farmer despises all other trades. If, like Augustine, one places happiness in the occupation with the highest good and makes the hereafter the criterion, then knowledge of worldly things is unimportant. With mystics like Molinos, this even goes so far that one's own self is devalued in the face of the more important divine truth. Humanists like Erasmus of Rotterdam see this quite differently and make human existence the center of interest. Following this and Augustine, Pascal advises focusing on one's own self and rejects pleasures and distractions as diversions.

Provoking curiosity, on the other hand, is the literary recipe of Mexía, when he can arouse the interest of his readers with wonderful individual cases and exceptional phenomena. In contrast, those who warn against useless books with lying stories and criticize curiosity and interest, reject the questioning of the why, Latin *cur*, of the *cur-ioso* as idleness. Villalón associates curiosity with desire and compares the thirst for knowledge with the greed for wealth. In the case of the Licenciado in the novel by Cervantes, wonder becomes the basic attitude in the face of the many unknown things he sees on his travels. After his transformation into a person who seems to know everything, curiosity is on the side of those who ask him idle questions. Equally idle and irrelevant are the unrealistic and speculative questions that Lazarillo is asked by the rector of the University of Salamanca. In his answers, Lazarillo exposes them and thus the university thought patterns as irrelevant. The criteria for the relevance of knowledge are often negotiated in literature in the early modern period.

For positivists and empiricists, there is no **doubt** that the object at hand should be the subject of scientific analysis. For the skeptics of the early modern period, it is not so simple. They question the prerequisites of knowledge in the subject, the object, and in intersubjective communication, and come to the conclusion that there are numerous, often insurmountable obstacles. Their critical attitude towards knowledge thus clearly differs from the attitude of positivists and empiricists, which would have seemed naive to them. The fact that what is far away like the stars and moves quickly, but appears slow to the senses, is for Sánchez evidence that the senses deceive. Different sensory organs in animals and humans or a missing sensory organ in a human according to Montaigne lead to the fact that we only know appearances, not the things as they are. Not least for this reason, opinions on a matter also vary among different people and at different times for the same person.

Gracián distinguishes between opinions and knowledge. While the former are believed to be correct, the latter are results of rational knowledge. Both can coexist when reason admits its limits and leaves certain areas to belief. Thus, dogmatism and skepticism are compatible. They were already in the Middle Ages, when people spoke of the double truth and distinguished between the truth of reason and the truth of revelation. In this way, the truths of the Bible apply, even if and because they are not checked by reason. With Gracián, this leads to him imparting numerous pieces of advice and insights dogmatically on the one hand, but on the other hand being skeptical about the limits of rational knowledge. If complete knowledge of the world were already realized, it would be possible for humans to recreate it at least in part. With Gracián, it requires the fictional and hypothetical figure of a *Veedor de todo*, if someone is able to recognize the substance of things along with the accidents. Normally, reason is clouded by many factors. Thus, each age of life has its forms of thought. The will can go beyond the ability of the mind. Emotion-driven, i.e., unreasonable interests in knowledge guide the mind. In the case of unclear truth options, one has to be content with probability. And if truth is finally what the competent experts consider it to be, then it must be conveyed through reading or social interaction. Gracián sees the main problem right here.

At court, truths are concealed or kept silent, as one does not want to be seen through by others. Complaisance is more important than truth. Even in the education of Andrenio by Critilo, they talk past each other. But above all, in the face of the great ignorant crowd, the wise Critilo deliberately holds back and keeps his knowledge to himself. He even presents himself as ignorant, as he would not be understood anyway. A main point of Gracián's skepticism refers to the communicability of knowledge, which is all the more important as the truth is initially in the possession of the wise and the competent experts.

Montaigne and Gracián thus draw attention to the idea that the object at hand is only an appearance, whose knowledge is clouded by senses, shaped by subjective predispositions, guided by affects and interests, gained from opinions and probabilities, and whose communicability is not unproblematic. Designating the empirically perceptible object of the natural sciences as positive, as the positivists of the 19th century did, would seem naive to the authors of the early modern period.

It can happen that something is known for sure and one is particularly proud of this knowledge. But if it turns out to be **pseudo-knowledge**, it is often made the subject of satire in literature. If the grammarian masters the verb forms and therefore considers himself competent to call Plato confused, if the historian knows a few numbers and details and thinks he can therefore judge the motives of the rulers' decisions, then both are exceeding their competence methodologically speaking. Their knowledge is useless where it matters. And if the arithmetician arrogantly wants to apply his knowledge of numbers to all disciplines or if the lawyer, referring to natural law, is mainly interested in demanding money, then both are driven by passions and miss the essential. Knowledge that is based on methodologically or morally questionable grounds is nothing more than pseudo-knowledge, which makes its representatives the subject of ridicule and literary satire.

If the natural sciences of the 19th century define nature as their subject, then man is a part of this nature. In the early modern period, he is the **crown of nature**. As a microcosm, he unites elements from all levels within himself. With his reason, he stands above nature, which has no reason, and as an actor, he can consciously set goals, where animate nature only has instincts. His free will allows him, according to Pico della Mirandola, the choice between an upward or downward development. Camos, referring to the Bible, assumes different degrees of perfection among living beings. The highest degree is found where reason is, thus in humans, who also distinguish themselves through their immortal soul. Finally, according to the Book of Genesis, the godlike man is to rule over the world. His expulsion from paradise is both a decline and an opportunity for development and perfection. Thus, man is not seen as just any part of nature, but stands above it with his characteristics, which the rest of nature does not have. From this perspective, literature that deals with man appears more valuable than natural science.

Epistemologically, man as a subject stands opposite the external world as an object. He is the one who recognizes this external world. Since he does this thinkingly, Descartes called him *res cogitans* and contrasted him with the external world, which he referred to as *res extensa*. Even if one does not see man in relation to the external world, but considers him as such, the material side of the body stands opposite the spiritual side, which was referred to as the soul in the early modern period, thus repeating the **contrast between external and internal** on an individual level. Here, as in the recognition of the world, the inner world is preferred in the early modern period, unlike the natural sciences in the 19th century. Already Plato had depicted the progress of knowledge in his allegory of the cave as a path from material appearances to ideas, from which the Spanish humanist Luis de León made a Christian pilgrimage. In the same sense, Augustine advises not to seek truth in the external material world, but within. Even when optics proceed geometrically or mathematically, it is subjective and internal tools that grasp the external space. If for Nicholas of Cusa, Descartes, and Pascal the soul and not the eye sees, then there is a devaluation, relativization, and subjectivization of the external world. Ignatius of Loyola, who speaks of *sentidos de la imaginación*, the senses of imagination, and Teresa of Ávila also agree with this. They share this opinion with Alejo de Venegas, for whom the inner senses are *ojos del alma*, or eyes of the soul. For Calderón de la Barca as for Luther, hearing is to be preferred to seeing, as it does not adhere to external appearance and is a more inwardly directed activity, which is not to be reduced to noise, but promotes primarily religious knowledge gain. The activities of the inner senses are rather to be situated in a *vita contemplativa* than in bustling world experience.

Action is goal-oriented. If one makes action the central category, then there are always also **goals** given, which are illustrated in fictional literature and treatises. The goal of man according to Aristotle is happiness. According to Plato, the creator's goal is that everything is good. According to Thomas Aquinas, animals are purposeful through their instinct. The goal thus becomes the *causa finalis*. Ethics, which deals with the right action, is always involved from this starting point. Goals become norms of action and their optimal achievement is considered perfection. There are boundary conditions that have a supportive effect. Thus, man as a social being achieves happiness only in a well-ordered state and not in a natural state. Against Aristotle, however, Hobbes defines man originally as wild and his goal, happiness, as a constantly progressing desire.

Goals and purposes were also found in medicine in the early modern period. Thus, according to Díaz, bones serve the purpose of supporting the rest of the body parts, and the eye is round for the purpose of being able to look everywhere. Moreover, and more importantly, the soul is seen as the purpose of the body. According to Suárez, the active soul is the goal and formal cause for the passively conceived body. If it is misguided, this has consequences for the body. It becomes sick. Therefore, according to Sabuco, negative affects of the soul cause physical diseases. For Merola, man is the final cause of the macrocosm and virtue is the origin and goal in the human microcosm as well as in the macrocosm. However, there is not always agreement regarding the goals. While Gassendi, as a follower of Epicureanism, defines the pleasures of the mind and body, even at the expense of others, as the goal, the Stoic Lipsius sees the goal in a state of mind unclouded by affects, which allows achieving the goal of happiness through virtue.

So, if action and its goals are made the starting point in different areas, then ethics is involved from the outset in every discipline of knowledge and does not come as an additional afterthought, as when the discipline is supplemented by the establishment of chairs for medical ethics or business ethics.

Human actions are subjects of literature. Goals of actions are chosen by free decision and thus become causes. This distinguishes them from the causal causes in natural science. So, if the starting point is freedom on the one hand and causal regularity on the other, then literature propagates the former and natural science the latter.

This approach is continued in the **German Romanticism** and in the philosophy of German Idealism. If for Schelling the divine will creates nature and thus gives predominance to the ideal over the real, and if he grants all living beings the freedom to realize themselves, then he sees actions and goals in nature and transfers the productive power of the absolute spirit to the individual spirit. This is followed by German physicists of the 19th century. Kastner understands nature as a manifestation of the divine spirit and defines the goal of the physicist as finding one's own laws of life in the life of the whole of nature. Since all organisms have a soul, natural history becomes the history of the development of the spirit. Like him, the physicist Eisenlohr also sees a spirit of order at work, achieving the most wonderful purposes with the simplest means. This shows how, even in the 19th century, literary guidelines of Romanticism and German Idealism led to a teleological understanding in physics.

Approach, realism, arrangement, and style are interconnected when Pascal distinguishes the *esprit géométrique* from the **esprit de finesse**. While the latter sees things spontaneously, intuitively, and directly, the former proceeds deductively, starting from general principles. Fontenelle wants to apply the *esprit géométrique* to as many areas as possible and, on the other hand, assigns the intuitive method to the individual phenomena of poetry and history. For Saint-Evremond, however, the *esprit géométrique* is incompatible with courtly manners, where entertainment and enjoyment are paramount. According to d'Alembert, courtly life would have distracted Descartes so much that he would not have been able to apply algebra to geometry. He himself wishes for short definitions for the encyclopedia, modeled on geometry, whose systematic approach he makes a model for all sciences. Diderot considers this outdated and prefers, in the sense of Pascal's *esprit de finesse*, a genius proceeding intuitively, similar to the courtly *honnête homme*. The controversy repeats itself with Linnaeus and Buffon. The former reduces the animal world to a few criteria, such as toes and feet, for the sake of his classifications, while the latter describes concrete phenomena in their entirety, complexity, and diversity based on individual observations. Thus, the intuitive engagement with individual phenomena seems to be more attributable to poetry, while the approach modeled on geometry appears scientific. However, the two methods can also work together, as Fontenelle's synthesis of the *poète philosophe* exemplifies. In any case, the controversy does not come to a clear conclusion, so neither the *esprit géométrique* nor the *esprit de finesse* should be preferred.

Usually, **rhetoric** is assigned to **literature** and **logic** to **natural science**, although scientific texts can also have rhetorical elements and literary texts should not be entirely without logic. Balzac uses theoretical elements from both logic and rhetoric, which for him is a kind of poetics. Moreover, he is a novelist who is inspired by scientific findings, without, however, simply adopting them. Instead, he takes the trouble to modify them when transferring them to the literary field. Like the zoologist Saint-Hilaire, he assumes the same structure in humans and animals, which is why there are analogies but also differences, as evidenced by the diverse manifestations of female existence in humans due to social conditions. Just as archaeologists draw conclusions about an entire environment from a single detail, so Balzac seeks to draw conclusions from a single chair to the entire furnishings.

With Saint-Hilaire, he also rejects the mathematical-geometrical style as unsuitable. Instead, he finds the rhetorical method of *amplificatio*, which derives further concrete sentences as implications from basic general sentences, to be sensible. In the writing process, the doctrine of the *loci communes* in rhetoric, as well as Aristotelian logic and the doctrine of definition by *genus proximum* and *differentia specifica*, can be helpful. The rhetorical *amplificatio* through the use of the *loci communes* was a common school exercise in 19th century France. In the rhetoric of praise, there was a whole register of question formulas, the answers to which could generate texts.

Balzac sees himself as an archaeologist of the social world, a classifier of professions, and a registrar of good and bad. This descriptive inventory is transformed into dynamics in the fictional novels. This is particularly demonstrated by the literary physiologies in France, which analyze and then satirically stage the customs of different professions or social groups through classification, deduction, and exemplary demonstration.

Despite all the references to the natural sciences popular with Balzac, the limits of natural science become clear again and again. They are evident, for example, in *La peau de chagrin*, where in the face of the *force vitale*, the leather skin, the biologist, the physicist, and the chemist fail with their analyses just as the doctors do in treating Raphael's disease. When the elements of natural science in Balzac are enriched with vitalism and metaphysics, and logical classification is supplemented by rhetorical amplification, literature and natural science seem to be in balance.

For the Spanish author of the realistic novel Pérez Galdós, literature, along with the freedom of the spirit and beauty in the arts, stands on the side of tradition. Opposite it is the technical-scientific **further development**. Both positions are presented with their advantages and disadvantages. If it was religious traditions that hindered scientific thinking in Spain and led to passivity, then for Galdós, openness and rational thinking are an antidote. However, from a social Darwinist perspective, Galdós shows that what is progress for one can lead to downfall for another, which is why he has ambivalent feelings towards progress.

In 17th century France, it was the ***Querelle des Anciens et des Modernes*** in which the advantages and disadvantages of tradition and progress were discussed. Initially about the exemplary effect of ancient literature compared to the present, it became a debate in the 19th century about the advantages and disadvantages of the time before and after the French Revolution, in which the defenders of the Ancien Régime and the advocates of progress and reason faced each other. While in the 17th century Perrault attributed a role model character to the age of Louis XIV because of its achievements in natural science, architecture, and painting, the scholar Boileau insisted on the superiority of ancient literature. Fontenelle, on the other hand, acknowledged the advances of the present in physics, medicine, and mathematics, although he doubted their usefulness for human happiness, and certified poetry to have produced everything essential very early on. In the 19th century, the debate was enriched by political implications when monarchists and counter-enlighteners rivaled liberals and socialists. While romantics like Chateaubriand considered the medieval Gothic cathedrals as wonders of nature, Zola's novels and Taine's literary criticisms were positivist. While in historiography Alexis de Tocqueville sought the reasons for the end of the Ancien Régime, Luis Blanc limited himself to positivist, uncommented facts. Thus, the *Querelle des Anciens et des Modernes* changed. Initially about the evaluation of literary role models, then about the value of reason and tradition, it was finally scientific methods, and their application in the consideration of history and literature, that were practiced and discussed.

If the evaluation is controversial in the *Querelle*, the question of precedence does not arise in the **cyclical thinking**, where the previous always returns, comparable to the change of seasons or day and night. And because progress is often associated with the natural sciences, they play no role in cyclical thinking. In antiquity, Hesiod distinguished ages, the golden, the silver, bronze, and iron, which cyclically alternated. For Polybius, the circularity of being becomes the inner reason for the rise and fall of states. Augustine did not see the Roman Empire fall, but in the sense of a *translatio imperii* pass into the Frankish Empire. While Voltaire lamented the cultural decline after the time of Louis XIV, the Enlightenment thinker Montesquieu saw himself at a new beginning, which was preceded by a decline.

The original human of a hypothetical **beginning time** has been imagined since the Middle Ages as living wild and irrational in the forest. How such a state can be overcome through education and upbringing is shown in novels like Gracián's *Criticón*. While Bodin, Vives, and Hobbes see the natural state characterized by egoism, envy, and conflicts and appreciate its overcoming, Catholics like Suárez and Protestants like Quenstedt argue about the possibility of a state of nature. The contribution of the sciences to overcoming the state of nature is differently assessed. While for Turgot the natural scientific developments appear beneficial for morality and manners, for Rousseau they are the cause of egoism, greed, and convenience. Technological progress thus appears as a regression for social skills.

The attribution of **progress** or decline is at the same time an evaluation. Progressive ideas are seen by French Enlightenment thinkers especially in England, where Bacon paved the way for experimental philosophy, Locke's empiricism appears as the preferred method, and Newton's scientific findings are exemplary. While the French Descartes and Pascal are considered outdated in the eyes of Voltaire, d'Alembert only appreciates their scientific and mathematical insights. D'Alembert's praise is for John Locke, who made metaphysics the experimental physics of the psyche. Progress is reflected in the inventions of new things. Since even currently common cultural goods were once invented, dealing with them since antiquity is a genealogy of traditions, in which needs acted as a culture-creating principle. Thus, the art of war was explained from the need of the weak good to defend themselves against the strong evil, which also legitimized the invention of gunpowder. According to Bacon, inventions do not happen by chance, but are results of a goal-oriented experimental experience controlled by the mind.

When technical achievements are held responsible for moral **decline**, Sparta is often used as a benchmark. According to Xenophon, Spartan education includes hardening by running without shoes and wearing the same cloak in summer as in winter. Even Mandeville saw the Spartans characterized by lack of need. This physical lack of need is praised by Rousseau and even before him by Seneca, when he presents the development of humanity from its beginnings to the present. In the past, people slept well on the hard ground; the dense forest protected them from sun, storm, and rain. When man had no house and no clothes, he was more resistant. Without an ax, his fist was strong. Without a sling, he could throw further, and without a horse, he could run faster. Technology, which softens through conveniences, is rejected.

Blacksmithing and agriculture bring further complications according to Rousseau. The blacksmith, who needs food but does not produce any, forces another to produce for two. Through division of labor, natural equality is lost. Agriculture leads to the division of agricultural land. With property, laws arise that secure inequality and wealth. This leads to the loss of good ways of life, in which reason and socialization contribute.

Voltaire sees the development in reverse. He reacts polemically to Rousseau's theses when he feels animated to walk on all fours again and live with the natives in Canada. Voltaire imagines the original state with Adam and Eve having long fingernails and disordered hair, which is why he considers scissors for hair and fingernails a sensible luxury. Technology can thus lead to luxury, against which Hume states two attitudes: A strict one, which strictly morally condemns the simplest luxury as the cause of corruption, and a liberal one, which sees the advantages of luxury for society. Building on this, Mandeville presents in his fable of the bees a flourishing state that is only condemned to decline after the disappearance of pride, luxury, and crime and fails due to lack of productivity.

Whether the conveniences that technology allows lead to the decay of morals and communities as Rousseau suggests, or whether they promote societal productivity, as Voltaire, Hume, and Mandeville imply, is thus controversial. Perhaps Seneca offers the solution when he distinguishes between natural desires, which have limits, and unnatural ones, which know no bounds. Aristotle too had recommended the middle measure, which lies between excess and deficiency. He criticized the Spartans for their one-sidedness. For their state system was only oriented towards martial abilities, which worked well as long as they were at war. But then, when they had achieved hegemony, they perished because they did not know how to live in leisure and were not practiced in any nobler virtue than the martial.[1]

The Englishman Francis Bacon also values technical achievements and inventions and expects them to improve human conditions and expand power over nature. When Voltaire elevated England as a model, it was primarily the methods of the natural sciences that he praised. Bacon wants to start from individual observations and derive general knowledge from these by **induction**, with the understanding that the knowledge gained through experiments must be processed in the mind. He rejects Aristotelian substances like final causes as anthropomorphism. For too long, people have been preoccupied with moral philosophy in antiquity and theology in the Middle Ages, neglecting the natural sciences. As with syllogistic deduction, the inductive method can also be applied to other sciences such as ethics and politics, with Mill noting that deduction is actually only a secondary procedure that fundamentally builds on inductively obtained propositions.

External perceptible objects are, according to Locke, the starting point of experience, which provides the material for reason. What the senses perceive becomes the material of thought, which assembles representations into ideas. Errors are also possible, for example, when ghosts are associated with darkness. Since for Hume there are no substances, the peach is nothing more than the composition of representations of taste, color, shape, size, and consistency. The same applies to the soul, whose representation is composed of perceptions, thoughts, and sensations and has no underlying substance. Hume adds the uniformity hypothesis, according to which regular empirical experience forms an expectation for the future. When repeating the experiment, the same effects are to be expected in space and time given the same causes.

When this approach is applied to morality, then the difference between the morally right and wrong is not fundamentally determined, but empirically to be explored. Here Smith introduces an impartial and well-informed spectator as a criterion of morality, into whose perspective one should put oneself and empirically judge one's own behavior. In this way, one experiences the beauty of nobility and the ugliness of injustice.

If the empiricists reject the Aristotelian concept of substance and accident, they also criticize everyday language, which suggests such a concept with its sentence parts subject, attribute, predicate, and object. According to Bacon, **language** should begin with definitions. It would then quickly become clear that there are names that do not correspond to anything in reality. He cites happiness or the prime mover as examples. Also problematic are names of things that do exist, but are confused. Here, Locke gives the example of "moistness", which sparked a dispute among scholars that ended when a clear definition was agreed upon. He considers figurative expressions, as they are commonly used in literature, to be particularly dangerous because they bring

---

[1] Aristotle 1995d, 65.

up false ideas and mislead. That the false use of language can create pseu-do-problems was already suggested by Herder. The Vienna Circle drew the conclusion from this that philosophy should deal with the difference between meaningful and meaningless sentences and leave statements about the object world to the individual sciences. Carnap therefore distinguishes sentences that say something about logical relationships from those that are empirically ver-ifiable. Following Bacon, he names a third group of words that claim to say something about reality but are meaningless, such as "world soul". Although the language criticism based on empiricism aims at scientific knowledge, it also affects literature, whose claim to truth it relativizes, and not only by rejecting the figurative and metaphorical mode of representation.

The subject of empirical observation is the immediately given. Therefore, certainty lies not as with Descartes in self-reflection on thinking, but in the material perceptible objects, the positively given facts. This is emphasized by Comte, who describes his basic attitude as positivist, and as a natural scientist prefers the actual, precise, and everyday to the imagined. Sociology becomes social physics for him, and psychology becomes phrenological physiology. If the sciences are hierarchically ordered, mathematics is at the top, while soci-ology is at the bottom due to greater complexity. In history as in the indi-vidual sciences, he sees a development from a theological to a metaphysical stage to a scientific or positive stage. He wants theological explanations of the world to be replaced by verifiable facts, which is why humanity as such should now be worshipped religiously. If Comte absolutizes the natural sciences and makes them a worldview, then questions of meaning and relevance, as posed by literature, are excluded.

When Bacon established the procedure of inferring causal laws from indi-vidual observations for the natural sciences, he also took a step further. He transfers his method to his understanding of nature, in which he sees only **causal laws** at work. This contradicts Aristotle, who also assumed final and efficient causes as well as substances. The rejection of teleological thinking, i.e., the assumption of immanent final and efficient causes, is generalized in the 19th century by Darwin and applied to the entire development of nature and man. This absolutization of the scientific method not only leads to ideo-logical consequences and controversies, but also makes goal-oriented human action, as it occurs in literature, secondary.

When Aristotle still asked the question of the purpose of existence and saw actions determined by a what for, for Haeckel man is an evolution prod-uct conditioned by natural causal processes. According to Lamarck, changes in environmental conditions were causes for changes in organisms, then in Darwin's theory of evolution, adaptation to the environment was the cause for the survival of organisms. Originally, there were a certain number of basic types in the animal kingdom from which the now existing species developed through natural and sexual selection. The latter ensures that in the competi-tion for mates, only the most beautiful and strongest prevail.

Opponents accuse Darwinism of starting from matter and not from spirit and therefore being materialistic. They refer to Aristotle and Thomas Aquinas, according to whom the main goal of animals is self-preservation while man strives for happiness. Attempts at mediation assume a transcendent act of world creation, after which the world is left to immanent goals, or they refer to the Aristotelian habitus, which results from repetitions and exercises and can lead to a new orientation or species. Since the stability of things cannot be explained without goal orientation, another attempt at mediation distinguishes between immanent tendency and transcendent goal. The principle of chance in mutations is opposed to the principle of necessity in goal orientation, from which the stability of things arises. If one derives a general historicization and relativization from the principle of chance, natural law also becomes invalid.

A consistently naturalistic literature, i.e., oriented towards the natural sciences, would therefore have to limit itself to observation, experiment, and causal laws. Purposeful human actions would be excluded. Naturalist Zola cannot go that far with his experimental novels, even if he claims this for himself. He aims to transfer the experimental method of the physician Bernard to literature by choosing characters for his novels whose actions are not determined by mental factors and whose motives and goals would not be understood in the sense of Montaigne, but are physically caused by blood and nerves and therefore causally explainable. Since the human body functions like a machine, it is possible to experimentally confirm or falsify hypotheses of laws, from which new hypotheses then arise. If the natural scientist has the same approach as the novel author, the novel becomes the experimental protocol of an experimenter. Zola wants to confirm the laws of inheritance and conditioning by the environment with his novels. This is only possible with inheritance when viewed over the long term. When Zola "observes" the history of the Rougon-Macquart family over five generations, he sees the hypothesis confirmed that hereditary defects and alcoholism lead to increasing degeneration. Initially, Zola was convinced that traits acquired during life are inherited, but he later saw this refuted by the Weismann barrier. Not least because of this, his last novel *Le Docteur Pascal* paints a less optimistic picture of positivist natural science.

The stories of the physician Ramón y Cajal promote scientific thinking like Zola's novels. However, since they are less realistic and belong more to the genre of science fiction, they are referred to as ***narraciones pseudocientíficas***. When a researcher arbitrarily administers tuberculosis bacteria to his rival and ages his too young wife with a drug so that she does not cheat on him again, or when a scientist would rather forego his newly acquired microscopic vision, scientific inventions play a role, but the protagonists pursue goals in their actions and are not subject to causal factors like in Zola. Nevertheless, Ramón y Cajal is convinced of the scientific methods when he makes observation, induction, and experiment the basis of knowledge and considers the introspection of a Descartes or Hegel a waste of time or calls for forgetting all book knowledge and thus tradition and history. His scientists are active and have no time for contemplation. He shares Comte's view that with advancing development, religion is replaced by natural science, and with social Darwinism, the idea of the further development of species depending on the requirements of the environment. The latter is evident in the soldiers who become extensions of their weapons. The fact that this scientifically shaped world is hostile to culture is shown by the disappearance of music and the novel, which has been replaced by the natural history of humanity.

The priority of the natural sciences is criticized by Ortega y Gasset. Nature and natural research are only one dimension of human life for him, whose actions have meaning and goals, and are therefore not to be understood causally, but teleologically. The life plan is a design for the future and does not originate from nature, but from history. The past belongs to personal identity, as does the horizon of historical circumstances. Therefore, there is no adaptation to circumstances as Darwin claims, but their conscious change through technology. Darwinism appears to Ortega, like the entire empirical natural science, no better than a myth that spreads imperialistically despite its gaps. Scientific specialization leads to the exclusion of challenging problems and to the informed ignorant, whose competence is limited to a small area and who rejects dealing with the entirety of knowledge as dilettantism. And even in a small scientific specialty, observation is not a pure reproduction of reality, but also the production of the observed, so that the image of reality is a conventionalist construct. This is shown when Euclidean geometry is replaced by Riemannian curves or quantum mechanics by the theory of relativity. Ortega uses the latter as the basis for his perspectivism, which sees humans in their respective historical situation. By subordinating the *razón pura* to the *razón vital*, nature appears subjectively shaped and loses its objective character. When he starts from human action, there are no causalities, but goals and plans. Since he holistically makes the context of historical and cultural circumstances the basis, individual scientific findings are nothing more than fragments.

In Borges, the criticism of the priority of the natural sciences is expressed in that metaphysical concepts and problems dominate in his **fantastic narratives**. If a memory that has all the details leads to the loss of thinking and abstraction abilities, if a classification disregards any logic and systematics, or if the letter of an alphabet becomes the point in which the entire universe is concentrated, then this is "fantastic" and confuses the reader. If finiteness and infinity are related to modes of locomotion as well as to infinite chains of variations of finite basic elements, then it is not truth, but astonishment that is conveyed. Fiction and reality become confused, where the rewriting of an already existing novel leads to an identical text that is better. Borges does not present solutions, but he provokes astonishment, which is not insignificant, as astonishment for Aristotle was the beginning of all philosophizing.

Bioy Casares also mixes fiction and reality in his narratives. Unlike Borges, he confronts natural science and technical inventions. If the fictional story of a film becomes reality, which the viewer can become a part of, then this is also fantastic, but it is closer to the genre of science fiction, which captures and shapes the advances and possibilities of the natural sciences in a literary way.

Taking a step further, Sábato sees the significance of literature in that it deals with metaphysical questions and the meaning of life. In contrast, he views scientific inventions as child's play. By using everyday language, literature conveys a complete picture of reality and not just parts of it like the scientific language, which has become so autonomous that it produces another type of fantastic literature. The mathematical-logical language leads to a loss of reality and the evaporation of the everyday world, as it, for example, cannot express a sense of justice. While natural science seeks general laws, literature moves in the concrete. It is oriented towards the subject and his lived experiences and feelings. While Ortega made life the central category, for Sábato it is the lived experience with its diverse subjective worldviews. Art and literature have more truths at hand about the latter than any statement about objective reality. Sábato also practices criticism of reason as language criticism, but in favor of everyday language, which is the language of literature, and at the expense of scientific language. This is particularly evident where he appreciates the metaphor, which particularly well expresses subjective worlds. When it equates unequal things, it operates similarly to literary fiction, whose actions stand for lived experiences and ideas. While Locke rejected the figurative expressions common in literature as misleading, for Sábato it is precisely these that can illuminate the deeper layers of reality.

REFERENCES

PRIMARY LITERATURE

Anónimo: *Segunda Parte del Lazarillo*. Pedro M. Piñero (ed.). Madrid: 1988.

Aristoteles: *Metaphysik*. Hamburg: Meiner 1995a.

Aristoteles: *Nikomachische Ethik*, Günther Bien (ed.). Hamburg: Meiner 1995b.

Aristoteles: *Physik*. Hamburg: Meiner 1995c.

Aristoteles: *Politik*. Eugen Rolfes (ed.). Hamburg: Meiner 1995d.

Aristoteles: *Rhetorik*. Gernot Krapinger (ed.). Stuttgart: Reclam 1999.

Aristoteles: *Über die Seele*. Hamburg: Meiner 1995e.

Augustinus: *Bekenntnisse*. Kurt Flasch (ed.). Stuttgart: Reclam 2000.

Augustinus: *De libero arbitrio. De vera religione*. Translated by Wilhelm Thimme. Zürich, Stuttgart: Artemis 1962.

Azpilcueta, Martìn de: *Manual de confessores y penitentes [...]*. Valladolid: Francisco Fernandez de Cordoua 1570.

Bacon, Francis: *Neues Organon*. Wolfgang Krohn (ed.). Hamburg: Meiner 1999, vol. 1.

Bacon, Francis: *Neues Organon*. Wolfgang Krohn (ed.). Hamburg: Meiner 2009, vol. 2.

Bacon, Roger: *Epistola de secretis operibus artis et naturae*. Hamburg: 1618.

Balzac, Honoré de: *Avant-propos*, in: Pierre-Georges Castex (ed.): La Comédie humaine. études de moeurs. Paris: Gallimard 1976, vol. 1, 7–24.

Balzac, Honoré de: *La peau de chagrin*. Pierre Barberis (ed.). Paris: Librairie générale française 1972.

Balzac, Honoré de: *Oeuvres diverses*. Paris: Éditions L. Conard 1938, vol. 2.

Balzac, Honoré de: *Physiologie du mariage*. M. Regard (ed.). Paris: Garnier-Flammarion 1968.

Bernard, Claude: *Introduction à l'étude de la médecine expérimentale* (1865). Paris: Garnier Flammarion 1966.

Bonniot, Joseph de: *La bête comparée à l'homme*. Paris: Retaux-Bray 1889.

C. Strosetzki, *Literature in Dialogue with the Natural Sciences*,
https://doi.org/10.1007/978-3-662-71319-8

Borges, Jorge Luis: *Nueva refutación del tiempo*, in: Otras inquisiciones, in: Obras completas. Buenos Aires: Emecé 1989–1996, vol. 2.

Borges, Jorge Luis: *Obras completas*. Buenos Aires: Emecé 1974.

Brillat-Savarin, Jean Anthelme: *Physiologie du goût*. Paris: Éditions Jouaust 1880.

Broglie, Abbé de: *La réaction contre le positivisme*. Paris: Plon 1894.

Bruyère, Jean de la: *Les caractères*. Robert Garapon (ed.). Paris: Garnier 1964.

Buffon, Georges Louis Le Clerc de: *Œuvres philosophiques de Buffon*. Jean Piveteau (ed.). Paris: Presses Universitaires de France 1954.

Buffon, Georges Louis Leclerc: *Discours, prononcé à l'Académie française, le jour de sa réception, le 25 août 1753*. Paris: 1904.

Cabriada, Juan de: *Carta filosófica, médico-chymica, en que se demuestra que de los tiempos, y experiencias se han aprendido los mejores remedios contra las enfermedades*. Madrid: Lucas Antonio de Bedmar y Baldivia 1686.

Cajal, Santiago Ramón y: *Los tónicos de la voluntad*. Madrid: Austral. 1971.

Cajal, Santiago Ramón y: *Obras literarias completas*. Madrid: Aguilar 1961.

Calderón de la Barca, Pedro: *Andrómeda y Perseo*. José Ruano de la Haza (ed.). Pamplona-Kassel: Universidad de Navarra–Reichenberger 1995.

Calderón de la Barca, Pedro: *El año santo en Madrid*. Ignacio Arellano/Carlos Mata (ed.). Pamplona–Kassel: Universidad de Navarra–Reichenberger 2005.

Calderón de la Barca, Pedro: *El nuevo palacio del Retiro*. Alan K.G. Paterson (ed.). Pamplona–Kassel: Universidad de Navarra–Reichenberger 1998.

Calderón de la Barca, Pedro: *El valle de la zarzuela*. Ignacio Arellano (ed.). Pamplona–Kassel: Universidad de Navarra–Reichenberger 2013.

Calderón de la Barca, Pedro: *El viático cordero*. Juan Manuel Escudero (ed.). Pamplona–Kassel: Universidad de Navarra–Reichenberger 2007.

Calderón de la Barca, Pedro: *La divina Filotea*. Luis Galván (ed.). Pamplona–Kassel: Universidad de Navarra–Reichenberger 2006.

Calderón de la Barca, Pedro: *La iglesia sitiada*. Beata Baczynska (ed.). Pamplona–Kassel: Universidad de Navarra–Reichenberger 2009.

Calderón de la Barca, Pedro: *La nave del mercader*. Ignacio Arellano con la colaboración de Blanca Oteiza/Mª Carmen Pinillos/Juan Manuel Escudero/Ana Armendáriz (ed.). Pamplona–Kassel: Universidad de Navarra–Reichenberger 1996.

Calderón de la Barca, Pedro: *La vida es sueño*. Ciriaco Morón (ed.). Madrid: Cátedra 1978.

Calderón de la Barca, Pedro: *Los encantos de la culpa. Estudio de Aurora Egido*. Juan Manuel Escudero (ed.). Pamplona–Kassel: Universidad de Navarra–Reichenberger 2004.

Camos, Antonio: *Microcosmia y gobierno universal del hombre cristiano, para todos los estados y cualquiera de ellos*. Madrid: Casa de la viuda de Alonso Gomez 1595.

Carnap, Rudolph: *Scheinprobleme in der Philosophie*. Frankfurt a.M.: Suhrkamp 1966.

Castrillo, Alonso de: *Tractado de república*. Madrid: Instituto de Estudios Políticos 1958.

Cervantes, Miguel de: *Don Quijote de la Mancha*. Francisco Rico (ed.). Barcelona: Instituto Cervantes, Crítica 1998.

Cervantes, Miguel de: *Novelas ejemplares II*. Juan Bautista Avalle-Arce (ed.). Madrid: 1982.

Cicero, Marcus Tullius: *De legibus / Über die Gesetze*. Rainer Nickel (ed.). Düsseldorf, Zürich: Artemis & Winkler 2002.

Comte, Auguste: *Discours sur l'esprit positif.* Paris: Carilian-Goeury 1844.

Corominas, Joan: *Breve diccionario etimológico de la lengua castellana.* Madrid: Editorial Gredos 1961.

Covarrubias, Sebastián de: *Tesoro de la Lengua Castellana o Española.* Ignacio Arellano/Rafael Zafra (ed.). Madrid: Iberoamericana 2006.

D'Alembert, Jean-Baptiste le Rond: *Discours préliminaire de l'Encyclopédie.* François Picavet (ed.). Paris: Armand Colin 1894.

D'Alembert, Jean-Baptiste Le Rond: *Œuvres complètes,* Genf: 1967a, vol. 1.

D'Alembert, Jean-Baptiste Le Rond: *Œuvres complètes,* Genf: 1967b, vol. 4.

D'Alembert, Jean-Baptiste Le Rond d': *Essai sur la société des gens de lettres et des grands,* in: Œuvres complètes, Genf: 1967c, vol. 4, 335–373.

Descartes, René: *Die Leidenschaften der Seele.* Klaus Hammacher (ed.). Hamburg: 1996.

Descartes, René: *La Dioptrique,* in: Descartes, Œuvres philosophiques 1618–1637. Ferdinand Alquier (ed.), Paris: Garnier 1988, vol. 1, 651–717.

Descartes, René: *Meditationen über die Grundlagen der Philosophie.* Lüder Gäbe (ed.). Hamburg: Meiner 1993.

Descartes, René: *Von der Methode.* Lüder Gäbe (ed.). Hamburg: Meiner 1971.

Diaz, Francisco: *Compendio de cirurgia y anatomia, en el qual se trata de todas las cosas tocantes a la theorica y pratica della, y de la anatomia del cuerpo humano, con otro breve tratado de las quatro enfermedades.* Madrid: Pedro Cossio 1575.

Diderot, Denis: *Premières notions sur les mathématiques à l'usage des enfants, dans Oeuvres complètes, Vol. II.* Paris: Hermann 1975.

Diderot, Denis: *Encyclopédie ou dictionnaire raisonné des sciences, des arts et des métiers.* Lausanne, Bern: 1781.

Diderot, Denis: *Œuvres philosophiques.* P. Vernière (ed.). Paris: Garnier 1964.

Eisenlohr, Wilhelm: *Lehrbuch der Physik zum Gebrauche bei Vorlesungen und zum Selbstunterricht.* Stuttgart: Krais und Hoffmann 1860, 8. edition.

Engels, Friedrich: Engels an Marx 11. oder 12.12.1859, in: *Marx-Engels-Werke,* Vol. 29, Institut für Marxismus-Leninismus beim Zentralkomitee der SED 1978.

Ermoni, Vincent: Finalisme et antifinalisme, in: *Annales de philosophie chrétienne 62,* 1892a, 346–371, 444–455, 481–506.

Ermoni, Vincent: Finalisme et antifinalisme, in: *Annales de philosophie chrétienne 63,* 1892–1893b, 53–83, 127–154.

Estella, Diego de: *Libro de la vanidad del mundo.* Madrid: Diputación Foral de Navarra y Editorial Franciscana Aránzazu 1980.

Fajardo, Diego de Saavedra: *República literaria.* Jorge García López Crítica (ed.). Barcelona: 2006.

Fetscher, Iring: Einleitung, in: *Auguste Comte: Rede über den Geist des Positivismus.* Hamburg: Meiner 1994.

Flaccus, Quintus Horatius: *Sermones et Epistulae.* Wilhelm Schöne (ed.). München: Ernst Heimeran 1953.

Fontenelle Bernard Le Bovier de: *Œuvres complètes.* Genf: Slatkine Reprints 1968a, vol. 1.

Fontenelle, Bernard Le Bovier de: *Œuvres complètes.* Genf: Slatkine Reprints 1968b, vol. 3.

Fontenelle, Bernard le Bovier de: *Oeuvres complètes.* Alain Niderst (ed.). Paris: Fayard 1989, vol. 3.

Fontenelle, Bernard le Bovier de: *Oeuvres complètes*. Alain Niderst (ed.). Paris: Fayard 1990, vol. 1.

Fontenelle, Bernard le Bovier de: *Oeuvres complètes*. Alain Niderst (ed.). Paris: Fayard 1991, vol. 2.

Galdós, Benito Pérez: *Doña perfecta*. Madrid: Casa editorial Hernando 1972.

Galdós, Benito Pérez: *Marianela*. Edu Robsy (ed.). Alayor: Maison Carrée 2016.

García Tapia, Nicolás: *Los veintiun libros de los ingenios y máquinas de Juanelo, atribuidos a Pedro Juan de Lastanosa*. Zaragoza: Departamento de Educación y Cultura 1997.

Gardeil, F. A.: *L'Évolutionnisme et les principes de S. Thomas*, in: Revue Thomiste. Questions du temps present. 1re Année, 1893a, 27–45, 316–327, 725–737.

Gardeil, F. A.: *L'Évolutionnisme et les principes de S. Thomas*, in: Revue Thomiste. Questions du temps present. 2ᵉ Année, Paris: Bureaux de la Revue Thomiste, 1894, 29–42.

Gardeil, F. A.: *L'Évolutionnisme et les principes de S. Thomas*, in: Revue Thomiste. Questions du temps present. 3ᵉ Année, Paris: Bureaux de la Revue Thomiste, 1895, 61–84, 607–633.

Gardeil, F. A.: *L'Évolutionnisme et les principes de S. Thomas*, in: Revue Thomiste. Questions du temps present. 4ᵉ Année, Paris: Bureaux de la Revue Thomiste, 1896, 29–42, 64–86, 215–247.

Gardeil, F. A.: *Notre Programme. Revue Thomiste. Questions du temps present.* Paris: P. L'ethielleux, Libraire-éditeur 1893b.

Gerson, Johannes: *Contra vanam curiositatem*, in: V. M. Glorieux (ed.): Œuvres complètes. Paris: 1962, vol. 3.

Goethe, Johann Wolfgang: *Geschichte der Farbenlehre*, in: Robert Boyle/Ernst Beutler (ed.): Gedenkausgabe der Werke, Briefe und Gespräche. Zürich: 1949.

Gracián, Baltasar: *El Criticón*, in: Arturo del Hoyo (ed.): Gracián, Obras completes. Biblioteca Castro 1993.

Gracián, Baltasar: *El Héroe, El Político, El Discreto, Oráculo manual y arte de prudencia*, in: Arturo del Hoyo (ed.): Oráculo manual. Barcelona: 1986.

Granada, Fray Luis de: *Introducción del símbolo de la Fé*. Buenos Aires: Espasa Calpe 1948.

Grimm, Jacob: *Kleinere Schriften: Vorreden, Zeitgeschichtliches und Persönliches*. Hildesheim, Olms: Weidmann 1966, vol. 8.

Heidegger, Martin: *Sein und Zeit*. Tübingen: Niemeyer 2006.

Herder, Johann Gottfried: *Verstand und Erfahrung. Vernunft und Sprache* (1799). Brüssel: 1969, vol. 2.

Hobbes, Thomas: *Leviathan*. Hermann Klenner (ed.). Hamburg: Meiner 1996.

Hobbes, Thomas: *Vom Menschen. Vom Bürger*. Günter Gawlick (ed.). Hamburg: Meiner 1966.

Homer: *Ilias. Odyssee*. Johann Heinrich Voß (ed.). München: Deutscher Taschenbuchverlag 2002.

Hume, David: *Abriß eines neuen Buches, betitelt: Ein Traktat über die menschliche Natur, etc.* (1740). Brief eines Edelmannes an seinen Freund (1745). Jens Kulenkampff (ed.). Hamburg: Meiner 1980.

Janet, Paul: *Les causes finales*. Paris: Germer 1876.

Jung, Carl Gustav: *Psychologie und Alchemie*. Zürich: Rascher 1944.

Laertius, Diogenes: *Lives of Eminent Philosophers*. Cambridge: 1950, Bd. 2.

Locke, John: *Versuch über den menschlichen Verstand*. Hamburg: Meiner 2006, vol. 1.

Locke, John: *Versuch über den menschlichen Verstand*. Hamburg: Meiner 2018, vol. 2.

Luther, Martin: *Predigt in Merseburg, 6. August 1545*, in: Weimarer Ausgabe 51, 11.

Luther, Martin: *Werke*. Weimar: 1883–2009.

Mauthner, Fritz: *Erinnerungen*. München: Georg Müller 1918.

Mauthner, Fritz: *Wörterbuch der Philosophie*. München, Leipzig: Georg Müller 1910.

Menéndez y Pelayo, Marcelino: *La ciencia Española*. Enrique Sánchez Reyes (ed.). Santander: CSIC 1953–1954a.

Menéndez y Pelayo, Marcelino: *La ciencia Española*, in: Enrique Sánchez Reyes (ed.). Santander: CSIC 1953–1954b, Bd. 2.

Merola, Hieronymo: *Republica original sacada del cuerpo humano*. Barcelona: Casa Pedro Malo 1587.

Mexía, Pedro: *Silva de varia lección*. Antonio Castro (ed.). Madrid: 1989a, vol. 1.

Mexía, Pedro: *Silva de varia lección*. Antonio Castro (ed.). Madrid: 1989b, vol. 2.

Mexía, Pedro: *Silva de varia lección*. Antonio Castro (ed.). Madrid: 1989c, vol. 3.

Mirandola, Pico della: *De hominis dignitate*. Gerd von der Gönne (ed.). Stuttgart: 1997.

Molino, Jean: *Quelques hypothèses sur la rhétorique au XIXe siècle*, in: Revue d'Histoire Littéraire de la France 79, 1980, 181–197.

Molinos, Miguel de: *Guía espiritual*. Madrid: Universidad pontífica de Salamanca 1976.

Mondragón, Jerónimo de: *Censura de la locura humana y excelencia Della*. Antonio Vilanova (ed.). Barcelona: 1953.

Montaigne, Michel de: *Œuvres complètes*. Maurice Rat (ed.). Paris: Gallimard 1962.

Montesquieu, Charles de Secondat: *De l'Esprit des Lois*. Robert Derathé (ed.). Paris: Garnier 1973.

Neufville, Etienne de: *Physiologie de la femme*. Paris, Aubert, Lavigne: 1842.

Ohm, Georg Simon: *Grundzüge der Physik als Compendium zu seinen Vorlesungen*. Nürnberg: J. L. Schrag 1854a, vol. 1.

Ohm, Georg Simon: *Grundzüge der Physik als Compendium zu seinen Vorlesungen*. Nürnberg: J. L. Schrag 1854b, vol. 3.

Oliva, Fernán Pérez de: *Diálogo de la dignidad del hombre*, J. L. Abellán (ed.): Ediciones de cultura popular. Barcelona: 1967.

Ortega y Gasset, José: *Historia como sistema*, in: Obras completas. Madrid: Revista de Occidente, 1964, vol. 6, 11–50.

Ortega y Gasset, José: *La rebelión de las masas*. Madrid: Revista de Occidente 1957, 163–171.

Ortega y Gasset, José: *Meditación de la técnica*. Madrid: Revista de Occidente 1939.

Ortega y Gasset, José: *Pasado y porvenir para el hombre actual*, in: Obras completas. Madrid Revista de Oriente, 1965, Bd. 9, 645–664.

Osuna, Francisco de: *Tercer abecedario espiritual*. Madrid: Biblioteca de autores cristianos 1972.

Pacheco, Luis: *Libro de las grandezas de la espada*. Madrid: Juan Iniguez de Lequerica 1605.

Pascal, Blaise: *Œuvres complètes*. Louis Lafuma (ed.). Paris: Editions du Seuil 1963.

Pascal, Blaise: *Pensées*. Philippe Sellier (ed.). Paris: Mercure de France 1976.

Pascal, Blaise: *Traité du vide*, in: Oeuvres. Léon Brunschvicg/Pierre Boutroux (ed.). Paris: 1908, vol. 2.

Perrault, Charles: *Parallèle des anciens et des modernes en ce qui regarde les arts et les sciences*. München: Eidos 1964.

Pineda, Juan de: *Diálogos familiares de agricultura*. Cristiana. J. Meseguer Fernández (ed.). Madrid: Atlas 1963, vol. 2.

Plinius secundus der Ältere: *Naturkunde*, Lateinisch/deutsch, book VII, Anthropologie. Roderich König (ed.). München: Heimeran 1975.

Popper, Karl: *Die offene Gesellschaft und ihre Feinde*. Tübingen: Mohr Siebeck 1992.

Pufendorf, Samuel von: *Acht Bücher vom Natur- und Völkerrecht*. Frankfurt: 1711.

Pufendorf, Samuel von: *De officio hominis et civis*. 1712.

Rigault, Hippolyte: *Histoire de la querelle des anciens et des modernes*. Paris: 1856.

Rousseau, Jean-Jacques: *Émile ou de l'éducation*. François Richard/Pierre Richard (ed.). Paris: Garnier 1964.

Sábato, Ernesto: *El escritor y sus fantasmas*. Buenos Aires: Aguilar 1971.

Sábato, Ernesto: *Hombres y engranajes. Heterodoxia*. Madrid: Alıanza Editorial 1988.

Sábato, Ernesto: *Uno y el universo. Edición definitiva*. Barcelona, México: Seix Barral 1981.

Sabuco de Nantes, Oliva: *Nueva filosofía de la naturaleza del hombre*. Atilano Martínez Tomé (ed.). Madrid: Editora Nacional 1981.

Saint-Evremond, Charles de Marguetel de Saint-Denis de: *Œuvres en prose*. R. Ternois (ed.). Paris: 1969a, vol. 2.

Saint-Evremond, Charles de Marguetel de Saint-Denis de: *Œuvres en prose*. R. Ternois (ed.). Paris: 1969b, vol. 4.

Saint-Evremond, Charles de Marguetel de Saint-Denis de: *Œuvres melees*. Amsterdam: Chez Pierre Mortier Libraire 1706, vol. 1.

Saint Hilaire, Isidore Geoffroy: *Histoire naturelles des regnes organiques*. Paris: Victor Masson 1854.

Saint-Hilaire, Geoffroy: *Notions synthétiques, historiques et physiologiques de philosophie naturelle*. Paris: Denain 1838.

Sánchez, Francisco: *Que nada se sabe*. Fernando A. Palacios (ed.). Madrid: Colección Austral 1991, 2. edition.

Scarion de Pauia, Bartolome: *Doctrina militar [...]*. Lissabon: Pedro Crasbeeck 1598.

Schelling, Friedrich Wilhelm Joseph: *Einleitung in die Philosophie*. Stuttgart: Frommann-Holzboog 1989.

Schlegel, Friedrich: *Vorlesungen über Universalgeschichte (1805–1806)*, in: Jean-Jacques Anstett (ed.): Kritische Friedrich-Schlegel-Ausgabe. München: 1960, vol. 14.

Smith, Adam: *Theorie der ethischen Gefühle*. Walther Eckstein (ed.). Hamburg: Meiner 2004.

Suárez, Francisco: *Disputaciones Metafísicas*. Sergio Rábalde Romeo/Salvador Caballero Sánchez/Antonio Pigcerver Zanón (ed.). Madrid: Gredos 1960–1966.

Suárez, Francisco: *Abhandlung über die Gesetze und Gott, den Gesetzgeber*. Freiburg, Berlin: Haufe 2002.

Suárez, Francisco: *De anima*. Salvador Castellote (ed.). Madrid: Sociedad de estudios y publicaciones 1978, vol. 1.

Taine, Hippolyte: *Les philosophes classiques du XIXᵉ siècle*. Paris: 1888.

Thionville, Eugène: *La théorie des lieux communs dans les topiques d'Aristote et les principales modifications qu'elle a subies jusqu'à nos jours*. Paris: 1855.

Thomas von Aquin: *Über die Herrschaft der Fürsten*. Stuttgart: Reclam 1971.

Thomas von Aquin: *De veritate catholicae fidei contra gentiles, libri quatuor.* Luxemburg: Brück 1881.

Thomas von Aquin: *In decem libros ethicorum Aristotelis ad Nicomachum expositio.* Rom: Marietti 1949.

Torquemada, Antonio de: *Jardín de flores curiosas.* Giovanni Allegra (ed.). Madrid: clásicos castalia 1982.

Valencia, Pedro de: *Obras completes.* Juan Francisco Domínguez (ed.). León: Universidad de León 2006, vol. 3.

Vico, Gian Battista: *De nostri temporis studiorum ratione. Vom Wesen und Weg der geistigen Bildung.* Darmstadt: Wissenschaftliche Buchgesellschaft 1984.

Villalón, Cristóbal de: *El scholástico.* Richard J.A. Kerr (ed.). Madrid: 1967.

Vives, Juan Luis: *De concordia et discordia, en: Opera omnia.* G. Mayans y Siscar (ed.), Valencia 1872–1890, vol. 5.

Vives, Juan Luis: *De disciplinis, in: Opera omnia in duos distincta tomos.* Basel: Nicolas l'Evesque 1555.

Vives, Juan Luis: *Tratado del alma.* Buenos Aires, Mexiko: Espasa-Calpe 1945.

Vives, Juan Luis: *Über die Gründe des Verfalls der Künste. De causis corruptarum artium.* Emilio Hidalgo-Serna (ed.). München: Fink 1990.

Voltaire: *Lettres philosophiques*, Raymond Naves (ed.), Paris: Garnier 1964.

Voltaire: *Œuvres complètes de Voltaire.* Paris: Imprimerie de la société littéraire-typographique 1785.

Wittgenstein, Ludwig: *Schriften.* Frankfurt a.M.: Suhrkamp 1969, Bd. 1.

Zola, Emile: *Le docteur Pascal.* Paris: Charpentier 1893.

Zola, Emile: *Le roman expérimental.* Aimé Guedj (ed.). Paris: Garnier Flammarion 1971.

RESEARCH LITERATURE

Altervogt, Heinrich: *Der Bildungbegriff im Wortschatze Ciceros.* Emsdetten: 1940 (Diss.).

Back, Mitja/Echterhoff, Gerald/Müller, Olaf, et al.: *Von Verteidigern und Entdeckern. Ein neuer Identitätskonflikt in Europa.* Berlin: Springer 2022.

Barceló, Rafael Ramís: *La segunda escolástica. Una propuesta de síntesis histórica.* Madrid: Dykinson 2024.

Baere, Benoît de: Représentation et visualisation dans L'Histoire naturelle de Buffon, in: *Dix-huitième siècle* 39, 2007, 613–638.

Becker, Colette: Aux sources du naturalisme Zolien 1850–1865, in: Pierre Cogny (ed.): *Le Naturalisme.* Paris: Hermann Éditeurs 2014, 9–33.

Behler, Ernst: Avicennas Hayy ibn Yakzan als Ausdruck des mittelalterlichen Platonismus, in: Kurt Flasch (ed.): *Parusia. Festgabe für Johannes Hirschberger.* Frankfurt: Minerva 1965, 351–375.

Bell, Tyler E.: *Galdós and Darwin.* London: Tamesis 2006.

Belting, Hans: *Florenz und Bagdad. Eine westöstliche Geschichte des Blicks.* München: C.H. Beck 2008.

Benavides, Manuel: Borges y la metafísica, in: *Cuadernos Hispanoamericanos* 505/507, 1992, 247–268.

Bender, Niklas: Das Motiv der Eisenbahn im Naturalismus – von der Technik zum Trieb (Zola, Hauptmann, Norris), in: *Poetica*, vol. 43, No. 1/2, 2011, 103–126.

Bender, Niklas: *Kampf der Paradigmen. Die Literatur zwischen Geschichte, Biologie und Medizin: Flaubert, Zola, Fontane*. Heidelberg: Winter 2009, 201–206.

Bernheimer, Richard: *Wild Men in the Middle Ages*. Cambridge: Harvard University Press 1952.

Bien, Günther: Zum Thema des Naturstands im 17. und 18. Jahrhundert, in: *Archiv für Begriffsgeschichte*, vol. 15, 1971, 275–298.

Bies, Michael: Naturwissen, natürlich. Die Méthode naturelle bei Buffon und Adanson, in: Silke Förschler/Nina Hahne (ed.): *Methoden der Aufklärung*. Brill 2012, 209–221.

Biesbrock, Hans-Rüdiger von: *Die literarische Mode der Physiologien in Frankreich (1840–1842)*. Frankfurt: 1978.

Binns, Niall: Nicanor Parra y la guerrilla litearia. Descifrando „Advertencia al lector", in: *Cuadernos Hispanoamericanos* 537, 1995, 83–99.

Blüher, Karl Alfred: Postmodernidad e intertextualidad en la obra de Jorge Luis Borges, in: Karl Alfred Blüher/Alfonso de Toro (ed.): *Jorge Luis Borges. Variaciones interpretativas sobre sus procedimientos literarios y bases epistemológicas*. Frankfurt, Madrid: Vervuert 1995, 119–132.

Blumenberg, Hans: *Der Prozeß der theoretischen Neugierde*. Frankfurt: 1973.

Blumenberg, Hans: *Die Lesbarkeit der Welt*. Frankfurt a.M.: 1981.

Bohrer, Karl Heinz/Scheel, Kurt: Zu diesem Heft, in: *Merkur* 8/9, 2007, 61, 657–658.

Borel, Jacques: *Médecine et psychiatrie Balzaciennes*. Paris: Corti 1973.

Borinski, Karl: *Baltasar Gracián und die Hofliteratur in Deutschland*. Halle/Saale: Niemeyer 1894.

Bosshard, Peter: *Die Beziehungen zwischen Rousseaus zweitem Discours und dem 90. Brief von Seneca*. Zürich: Juris Druck 1967 (Diss.).

Brieskorn, Norbert: *Rechtsphilosophie*. Stuttgart, Berlin, Köln: Kohlhammer 1990.

Brüllmann, Philipp: *Die Theorie des Guten in Aristoteles' Nikomachischer Ethik*. Berlin: De Gruyter 2011.

Bruns, Günter: Round-Table-Discussion: Zufall und Notwendigkeit in der Evolution, in: Joachim-Hermann Scharf (ed.): *Evolution*. Halle/Saale: Deutsche Akademie der Naturforscher Leopoldina 1975, 395–416.

Buck, August: *Die humanistische Tradition in der Romania*. Bad Homburg: Gehlen Verlag 1968.

Buck, August: *Die Rezeption der Antike*. Hauswedell Verlag 1981a.

Buck, August: Juan Luís Vives: Konzeption des humanistischen Gelehrten, in: August Back (ed.): *Juan Luis Vives*. Hamburg: Hauswedell 1981b.

Busche, Hubertus: Teleologie, teleologisch, in: Joachim Ritter/Karlfried Gründer (ed.): *Historisches Wörterbuch der Philosophie*. Darmstadt: Wissenschaftliche Buchgesellschaft 1998, vol. 10, 970–979.

Calderón, Ángeles García: El análisis de la ciencia por medio de la literatura: Newton interpretado por Voltaire, in: *Estudios Franco-Alemanes* 7, 2015, 19–36.

Campa, Riccardo: La comprensión como ficción, in: *Cuadernos Hispanoamericanos* 391/393, 1983, 7–34.

Canseco, Luis Gómez: *El humanismo después de 1600: Pedro de Valencia*. Sevilla: Secretariado de Publicaciones de la Universidad de Sevilla 1993.

Carpotta, James: *Kingship and Tyranny in the Theater of Guillén de Castro*. London: Tamesis Books 1984.

Cersosimo, Emilse Beatriz: De los carácteres a la metafísica, in: *Revista Iberoamericana* 58, 1992, 193–206.

Chavarri, Raúl: La metafísica y las metafísicas de Ernesto Sábato, in: *Cuadernos Hispanoamericanos* 391/393, 1983, 675-680.

Comparone, Loredana: Of Bacteria, Scientists, and Women: Ramón y Cajal's „El pesimista corregido", in: *Letras Femeninas* 43, 1, Summer-Fall 2017, 46–59.

Coy, Wolfgang: Eine Einheit der Wissenschaften, in: *Tagesspiegel-Sonderseiten: Humboldt-Universität*, 14.4.2003, 1–2.

Curtius, Ernst Robert: *Europäische Literatur und lateinisches Mittelalter*. Bern, München: Francke Verlag 1973.

Dapía, Silvia G.: *Die Rezeption der Sprachkritik Fritz Mautherns im Werk von Jorge Luis Borges*. Köln, Weimar, Wien: Böhlau 1993.

Darge, Rolf: Suárez' Metaphysikentwurf in der Geschichte des metaphysischen Denkens, in: Cornelius Zehetner (ed.): *Menschenrechte und Metaphysik. Beiträge zu Francisco Súarez*. Göttingen: V&R unipress 2019, 21–46.

Darge, Rolf: Zum historischen Hintergrund der Transzendentalienlehre in den Disputationes metaphysicae, in: Lukás Novák (ed.): *Suárez's Metaphysics in Its Historical and Systematic Context*. Berlin, Boston: De Gruyter 2014, 39–62.

Daston, Lorraine: Die Lust an der Neugier in der frühneuzeitlichen Wissenschaft, in: Klaus Krüger (ed.): *Curiositas*. Göttingen: 2002, 147–175.

Daus, Ronald: *Zola und der französische Naturalismus*. Stuttgart: Metzler 1976.

Davis, Ryan A.: Modern Spain, a Myth: Regeneration Thought Reeducation in Santiago Ramón y Cajal's Cuentos de vacaciones (1905), in: *Revista de Estudios Hispánicos*, XLVII, 2, Junio 2013, 313–335.

Delgado, Manuel: *Tiranía y derecho de resistencia en el teatro de Guillén de Castro*. Barcelona: Puvill 1984.

Disselkamp, Martin: Parameter der Antiqui-Moderni-Thematik in der Frühen Neuzeit, in: Herbert Jaumann (ed.): *Diskurse der Gelehrtenkultur in der Frühen Neuzeit*. Berlin: De Gruyter, 157–177.

Dreitzel, Horst: Zur Entwicklung und Eigenart der „eklektischen Philosophie", in: *Zeitschrift für historische Forschung*, vol. 18, 1991, 281–343.

Echevarría, Arturo: *Lengua y literatura de Borges*. Barcelona: Ariel 1983.

Eco, Umberto: *Die Suche nach der vollkommenen Sprache*. München: dtv Verlagsgesellschaft 2002.

Egido, Aurora: 'El Criticón' y la retórica del silencio, in: Sebastian Neumeister/ Dietrich Briesemeister (ed.): *El mundo de Gracián. Actas del Coloquio Internacional Berlín 1988*. Berlin: 1991, 13–20.

Egido, Aurora: *Las humanidades y Gracián*, in: ABC Cultural (31.03.2001).

Eisler, Rudolf: *Der Zweck. Seine Bedeutung für Natur und Geist*. Berlin: Mittler und Sohn 1914.

Eley, Lothar: Auguste Comte, in: Margot Fleischer/Jochem Hennigfeld (ed.): *Philosophen des 19. Jahrhunderts*. Darmstadt: Primus 1998, 144–159.

Engelhardt, Dietrich von: Medizin der Romantik, in: Werner E. Gerabek/Bernhard D. Haage/Gundolf Keil/Wolfgang Wegner (ed.): *Enzyklopädie Medizingeschichte*. Berlin, New York: De Gruyter 2005, 903–907.

Engelhardt, Dietrich von: *Medizin in Romantik und Idealismus*. Stuttgart: Frommann-Holzboog 2023.

Ertz, Stefanie: *Vertrag und Gesetz: das Naturrecht und die Bibel bei Grotius, Hobbes, Spinoza.* Würzburg: Königshausen u. Neumann 2014.

Eusterschulte, Anne: *Analogia entis seu mentis. Analogie als erkenntnistheorethisches Prinzip in der Philosophie Giordano Brunos.* Würzburg: Königshausen 1997.

Fernández, Teodosio: Ernesto Sábato y la literatura como indagación, in: *Cuadernos Hispanoamericanos 391/393,* 1983, 35–45.

Fetscher, Iring: Der gesellschaftliche „Naturzustand" und das Menschenbild bei Hobbes, Pufendorf, Cumberland und Rousseau, in: *Schmollers Jahrbuch für Gesetzgebung, Verwaltung und Volkswirtschaft* 80(1), Halbbd.: 1960, 641–685.

Filmer, Fridtjof von: *Das Gewissen als Argument im Recht.* Berlin: Duncker & Humblot 2000.

Fisac, Miguel Ángel Quintanilla: La ciencia y la cultura científica, in: *ArtefaCToS,* vol. 3, N.1, Diciembre 2010, 31–48.

Fischer, Peter: Zur Genealogie der Technikphilosophie, in: Peter Fischer (ed.): *Technikphilosophie.* Leipzig: Reclam 1996, 255–335.

Flasche, Hans: Ideas agustinianas en la obra de Calderón, in: *Bulletin de Hispanic Studies* 61, 1984, 335–342.

Fleming, Leonor: Creación y teoría literaria a propósito de ‚Las ruinas circulares', in: *Anthropos* 142/143, 1993, 115–118.

Föcking, Marc: *Pathologia litteralis, Erzählte Wissenschaft und wissenschaftliches Erzählen im französischen 19. Jahrhundert.* Tübingen: Narr 2002.

Foucault, Michel: *Die Ordnung der Dinge. Eine Archäologie der Humanwissenschaften.* Frankfurt: Suhrkamp 1974.

Foucault, Michel: *L'archéologie du savoir.* Paris: Gallimard 1969.

Foucault, Michel: *Schriften in vier Bänden.* Daniel Defert (ed.). Frankfurt a.M.: Suhrkamp 2002, vol. 2.

Fox, Dian: *Kings in Calderón: A Study in Characterisation and Political Theory.* London: Tamesis 1986.

Friedrich, Hugo: *Montaigne.* Bern, München: Francke 1967.

Frutos, Eugenio: *La filosofía de Calderón en sus autos sacramentales.* Zaragoza: Institución Fernando El Católico de la Excma 1952.

Fumaroli, Marc: *La Querelle des anciens et des modernes.* Paris: 2001.

Gadamer, Hans-Georg: *Wahrheit und Methode.* Tübingen: Mohr 1972.

Gambin, Felice: Anotaciones sobre el concepto de ‚virtud', in: Jorge M. Ayala (ed.): *Baltasar Gracián. Selección de estudios, investigación actual y documentación.* Barcelona: Anthropos 1993, 62–67.

Gärtner, Wolfgang: *Philosophische Impulse: Recht und Gerechtigkeit von Aristoteles bis Habermas: eine Einführung in die Ethik der Gerechtigkeit.* Berlin: LIT 2012.

Géruzez Eugène: *Cours de littérature, rédigé d'après le programme pour le baccalauréat.* Paris: 1842.

Giammusso, Salvatore: Sprache der Macht und Macht der Sprache, in: *Germanisch-Romanische Monatsschrift* 43/3, 1993, 302–314.

Goodhart, David: *The Road to Somewhere. The Populist Revolt and the Future of Politics.* London: 2017.

Grassi, Ernesto: Die Unfehlbarkeit. Ein philosophisches Problem. Sprache und Schau, in: *Kerygma und Mythos* 6.6, 1975, 51–69.

Grassmann, Franz L.: *Die Schöpfungslehre des heiligen Augustinus und Darwins.* Regensburg: Manz 1889.

Grinda y Forner, José: *Las ciencias positivas en Calderón de la Barca*. Madrid: Montoya 1881.

Grugel-Pannier, Dorit: *Luxus: eine begriffs- und ideengeschichtliche Untersuchung unter besonderer Berücksichtigung von Bernard Mandeville*. Frankfurt: Lang 1996.

Guderian, Myung Hee: *Perspektiven der Metaphysikkritik*. Paderborn: Mentis 2009.

Guillo, Dominique: À la recherche des signes de l'identité. Balzac et l'histoire naturelle, in: *Politix* 2006, vol. 2, Nr. 74, 49–74.

Gusdorf, Georges: *Introduction aux sciences humaines*. Paris: Les Éditions Ophrys 1974.

Gutiérrez Girardot, Rafael: Crítica literaria y filosofía en Jorge Luis Borges, in: *Cuadernos Hispanoamericanos* 505/507, 1992, 279–297.

Hajduk, Stefan: Experiment und Revolution – Zur ästhetischen Theorie des historischen Naturalismus, in: *Weimarer Beiträge* 51/2, 2005, 236–253.

Hanks, Lesley: *Buffon avant l' histoire naturelle*. Paris: Presses Universitaires de France 1966.

Harada, Eduardo: Einstein y Ortega: Relativismo. Teoría de la relatividad y perspectivismo, in: *Elementos* 62, 2006, 3–13.

Harkema, Leslie J.: Miau and Spanish Science, in: *MLN* 134/2, 2019, 266–285.

Häussler, Reinhard: Vom Ursprung und Wandel des Lebensaltervergleichs, in: *Hermes* 92, 1964, 313–341.

Heger, Klaus: *Baltasar Gracián. Estilo y Doctrina*. Zaragoza: 1982.

Hinz, Manfred: Mit der Wahrheit lügen. Baltasar Gracián und Juan Manuel, in: Joachim Küpper/Friedrich Wolfzettel (ed.): *Diskurse des Barock*. München: 2000, 15–46.

Hoffmann, Thomas Sören: Zweck, Ziel, in: Joachim Ritter/Karlfried Gründer/Gottfried Gabriel (ed.): *Historisches Wörterbuch der Philosophie*. Darmstadt: Wissenschaftliche Buchgesellschaft 2004, vol. 12, 1486–1510.

Holzhey, Helmut: Philosophie als Eklektik, in: *Studia leibnitiana*, 15, 1983, 19–29.

Huaulmé Kilian: Balzac et les géographes: reconversion d'un discours scientifique, in: *Studia Romanica Posnaniensia* 50, 1, 2023, 7–19.

Iriarte, Joaquin: *Kartesischer oder Sanchezischer Zweifel. Ein kritischer und philosophischer Vergleich zwischen dem Kartesischen ‚Discours de la Méthode‘ und dem Sanchezischen ‚Quod Nihil Scitur‘*. Bottrop 1935.

Jansen, Hellmut: *Die Grundbegriffe des Baltasar Graciáns*. Genf: E. Droz 1958.

Jolibert, Bernard: Science et religion chez Auguste Comte, in: *Expressions* 23, 2004, 105–121.

Joly, Robert: Curiositas, in: *L'Antiquité Classique* 30, 1961, 33–44.

Juárez, Alfonso Moraleja: *Baltasar Gracián: forma política y contenido ético*. Madrid: 1999.

Kasner, Norberto M.: Metafísica y soledad. Un estudio de la novelística de Ernesto Sábato, in: *Revista Iberoamericana* 58, 1992, 105–113.

Kaufhold, Martin: *Europas Werte*. Paderborn, München: Schöningh 2013.

Klingner, Friedrich: *Römische Geisteswelt*. München: Ellermann 1965, 5. Aufl.

Klinkert, Thomas: *Epistemologische Fiktionen. Zur Interferenz von Literatur und Wissenschaft seit der Aufklärung*. Berlin: De Gruyter 2010.

Klinkert, Thomas: *Fiktion, Wissen, Gedächtnis*. Baden-Baden: Rombach 2020.

Kluxen, Wolfgang: *„Lex naturalis“ bei Thomas von Aquin*. Wiesbaden: Westdt. Verl. 2001.

Köck, Heribert Franz: *Der Beitrag der Schule von Salamanca zur Entwicklung der Lehre von den Grundrechten.* Berlin: Duncker & Humblot 1987.

Konersmann, Ralf: Sehen, in: Joachim Ritter (ed.): *Historisches Wörterbuch der Philosophie.* Basel: Schwabe 1995, vol. 9, 121–149.

Kortum, Hans/Perrault, Charles/Boileau, Nicolas: *Der Antike-Streit im Zeitalter der klassischen französischen Literatur.* Berlin: Rütten & Loening 1966.

Krauss, Werner: *Graciáns Lebenslehre.* Frankfurt: Klostermann 1947.

Krauss, Werner: *Fontenelle und die Aufklärung.* München: Fink 1969.

Krüger, Klaus: Einleitung, in: Klaus Krüger (ed.): *Curiositas.* Göttingen: Wallstein 2002.

Kuener, Marianne: *Literatur und Philosophie: ihr Verhältnis bei Ernesto Sábato.* Frankfurt a.M.: 1991.

Kühn, Joachim: *Gescheiterte Sprachkritik, Fritz Mauthners Leben und Werk.* Berlin: De Gruyter 1975.

Küpper, Joachim: *Balzac und der effet de réel.* Amsterdam: BR Grüner 1986.

Lasinger, Wolfgang: *Aphoristik und Intertextualität bei Baltasar Gracián.* Tübingen: Gunter Narr Verlag 2000.

Lauer, Robert: Bandos y tumultos en el teatro político del *Siglo de Oro* (1994), in: Carmen Hernández (ed.): *Teatro, historia y sociedad.* Murcia, Editum 1996.

Lenoble, Robert: *Mersenne ou la naissance du mécanisme.* Paris: Vrin 1943.

Lettow, Susanne: Naturgeschichte und Geschichte der Menschheit, in: Johannes Rohbeck/Lieselotte Steinbrügge (ed.): *Jean-Jacques Rousseau: Die beiden Diskurse zur Zivilisationskritik.* Berlin: De Gruyter, 2015, 83–103.

Lind, Gunter: *Physik im Lehrbuch. 1700–1850.* Berlin, Heidelberg: Springer 1992.

Lovejoy, Arthur O.: Der vermeintliche Primitivismus von Rousseaus Abhandlung über die Ungleichheit, in: *Deutsche Zeitschrift für Philosophie*, Akademie Verlag, vol. 60, nr. 4, 2012, 491–508.

Löw, Reinhard: Evolutionstheorie als Weltanschauung und die Wirklichkeit der Welt, in: *Imago Hominis 1(2,5)*, 1994, 117–125.

Löwith, Karl: *Heilsgeschichte und Weltgeschehen.* Stuttgart: 1961.

Lubac, Henri de: *Exégèse médiévale. Les quatre sens de l'Écriture.* Paris: Aubier 1959, Bd. 2.

Lübbe, Heinrich: Staat und Zivilreligion, in: N. Achterberg (ed.): *Legitimation des modernen Staates.* Wiesbaden: Steiner 1981, 40–64.

Ludwig, Bernd: Lehrjahre im Exil? Zu einigen Wandlungen in Hobbes' Politischer Philosophie, in: Dieter Hüning (ed.): *Der lange Schatten des Leviathan.* Berlin: Duncker & Humblot 2005, 11–27.

Lühe, Astrid von der: Schmecken, in: Ralf Konersmann (ed.): *Wörterbuch der philosophischen Metaphern.* Darmstadt: Wissenschaftliche Buchgesellschaft 2007, 340–355.

Lüthe, Rudolf: John Stuart Mill, in: Margot Fleischer/Jochem Hennigfeld (ed.): *Philosophen des 19. Jahrhunderts.* Darmstadt: Primus 1998, 160–178.

Mahnke, Dietrich: *Leibniz als Gegner der Gelehrteneinseitigkeit.* Radebeul: Krause 1912.

Maravall, José Antonio: *Teatro y literatura en la sociedad barroca.* Madrid: Seminario y Ediciones 1972.

Margraff, Nicolaus: *Der Mensch und sein Seelenleben in den Autos Sacramentales des Don Pedro Calderón de la Barca.* Bonn: H. Ludwig 1912 (Inaugural-Dissertation).

Martinez, Guillermo: *Borges y la matemática*. Buenos Aires: Eudeba 2003.

Matzat, Wolfgang: Galdós und der französische Realismus/Naturalismus, in: Hans-Jürgen Lüsebrink/Hans T. Siepe (ed.): *Romanistische Komparatistik*. Frankfurt a.M.: Lang 1993, 127–145.

Mendoza, José María Felipe: Francisco Suárez y la física aristotélica. Notas sobre la filosofía de la naturaleza según las *Disputaciones Metafísicas* I, in: *Revista de Filosofía* 47, 1, 2022, 29–48.

Merrell, Floyd: *Unthinking Thinking: Jorge Luis Borges. Mathematics, and the New Physics*. West Lafayette: Purdue University Press 1991.

Mette, Hans Joachim: Curiositas, in: *Festschrift Bruno Snell zum 60. Geburtstag*. München: 1956.

Miquel, Mercedes: Filosofía de la Ciencia en Ortega y Gasset, in: *Logos. Anales del Seminario de Metafísica, Núm. Extra. Homenaje a S. Rábade*. Madrid: Ediciones Complutense 1992, 127–154.

Mordek, Hubert u.a. (ed.): *Die ‚Admonitio generalis' Karls des Großen*. Hannover: Hahn 2012.

Moreau, Joseph: Doute et Savoir chez Francisco Sanches, in: *Portugiesische Forschungen der Görresgesellschaft*. Münster: Aschendorff 1960, 24–50.

Moser, Simon: Die Änderung des Begriffes der Metaphysik bei Thomas von Aquin und Franz Suarez, in: Mors (ed.). *Metaphysik Einst und Jetzt*. Berlin: De Gruyter 1958, 13–32.

Müller, Jan-Dirk: Erfahrung zwischen Heilssorge, Selbsterkenntnis und Entdeckung des Kosmos, in: *Past & Present 73*, 1976, 307–341.

Müller, Reimar: Rousseau als Geschichtsphilosoph, in: *Sitzungsbericht der Leibniz-Sozietät der Wissenschaften zu Berlin* 117, 2013, 33–50.

Müller-Wille, Stefan: Genealogie, Naturgeschichte und Naturgesetz bei Linné und Buffon, in: K. Heck/B. Jahn (ed.): *Genealogie als Denkform in Mittelalter und Früher Neuzeit*. Tübingen: Max Niemeyer 2011, 109–119.

Myers, Richard: Christianity and Politics in Montesquieu's Greatness and Decline of the Romans, in: *Interpretation* 17/2, Winter 1989–90, 223–238.

Neffs, Jaques: La localisation des sciences, in: Claude Duchet (ed.): *Balzac et La peau de chagrin*. Paris: Société d'édition d'enseignement supérieur 1979, 127–142.

Neumeister, Sebastian: Gracián filósofo, in: Manuel García Martín et al. (ed.): *Estado actual de los estudios sobre el siglo de oro. Actas del II Congreso Internacional de Hispanistas del Siglo de Oro*. Salamanca: 1993, 735–739.

Neumeister, Sebastian: La observación del otro y de sí mismo en Gracián, in: *Ínsula* 655–656, 2001, 38–39.

Neumeister, Sebastian: Schopenhauer als Leser Graciáns, in: Sebastian Neumeister/ Dietrich Briesemeister (ed.): *El mundo de Gracián. Actas del Coloquio Internacional*. Berlin: 1988 (Berlin 1991), 261–277.

Noel, Valis: *Sacred Realism. Religion and Imagination in Modern Spanish Narrative*. New Haven, London: Yale UP 2010.

Nouailles, Bertrand: Le monstre: un concept stratégique dans l'Histoire naturelle de Buffon, in: *Revue philosophique de la France et de l'étranger* 1, Tome 141, 2016, 41–58.

Novillo-Corvalán, Patricia: Explorers of the Human Brain. The Neurological Insights of Borges and Ramón y Cajal, in: Patricia Novillo-Corvalán (ed.): *Latin American and Iberian Perspectives on Literature and Medicine*. New York: Routledge 2015, 23–44.

Parker, Alexander A.: Calderón, el dramaturgo de la escolástica, in: *RevEstHisp* 3, 1935, 273–285.

Parker, Alexander A.: Calderón, el dramaturgo de la escolástica, in: *RevEstHisp* 4, 1936, 393–420.

Perez, Andrew W.: Life in the Year 6000 by Santiago Ramón y Cajal: A Translation of an Unpublished „Vacation Story", in: *Literature and Medicine* 35, 1, Spring 2017, 203–228.

Pimenta, Pedro: Le „dessin originaire de la nature" dans l'*Histoire naturelle* de Buffon et Daubenton, in: *Dix-huitième siècle* 1, 2017, No. 49.

Pinto, Alba Milagro: Ortega y Gasset y la crítica de la rezón científica, in: *Brocar* 33, 2009, 195–223.

Plessner, Helmuth: *Die verspätete Nation.* Stuttgart: Kohlhammer 1959.

Pope, Stephen J.: Tradition und Erneuerung im Naturrecht. Eine tomasische Interpretation, in: *Concilium* 46, 2010, 253–282.

Popowicz, Kamil: Emile Zola entre Lamarck, Darwin et Weismann, in: Miroslaw Loba/Barbara Luczak (ed.): *Formes du vivant, formes de littérature.* Poznan: Naukowe 2019, 159–170.

Poppenberg, Gerhard: Ganz verteufelt human. Gracián als Moralist, in: Sebastian Neumeister/Dietrich Briesemeister (ed.): *El mundo de Gracián.* Berlin: 1988, 171– 202.

Pratt, Dale J.: *Signs of Science: Literature, Science, and Spanish Modernity since 1866.* West Lafayette: Purdue University Press 1992.

Pulte, Helmut: Wissenschaft, in: Annika Hand/Christian Bermes/Ulrich Dierse (ed.): *Schlüsselbegriffe der Philosophie des 19. Jahrhunderts.* Hamburg: Meiner 2015, 483–522.

Ramírez, Pedro M. Piñero: *Segunda Parte del Lazarillo.* Madrid: Cátedra 1988.

Recknagel, Dominik: *Einheit des Denkens trotz konfessioneller Spaltung. Parallelen zwischen den Rechtslehren von Francisco Suárez und Hugo Grotius.* Frankfurt am Main: Lang 2010.

Rehm, Michaela: Aufklärung über Fortschritt. Die systematischen Ursachen der Zivilisation, in: Johannes Rohbeck/Lieselotte Steinbrügge (ed.): *Jean-Jacques Rousseau. Die beiden Diskurse zur Zivilisationskritik.* Berlin: De Gruyter 2015, 47–60.

Reibstein, Ernst: *Völkerrecht. Eine Geschichte seiner Ideen in Lehre und Praxis.* Freiburg: Alber 1958, vol. 1.

Rest, Jaime: *El laberinto del universo. Borges y el pensamiento nominalista.* Buenos Aires: Ediciones Librerías Fausto 1976.

Riemann, Bernhard: Ueber die Hypothesen, welche der Geometrie zu Grunde liegen, in: *Abhandlungen der Königlichen Gesellschaft der Wissenschaften zu Göttingen.* 1876, Bd. 13.

Rivero, Carmen: *Humanismus, Utopie und Tragödie.* Berlin: De Gruyter 2020.

Scattola, Merio: *Das Naturrecht vor dem Naturrecht. Zur Geschichte des ‚ius naturae' im 16. Jahrhundert.* Tübingen: Niemeyer 1999.

Schalk, Fritz: Erasmus und die res publica literaria, in: *Akten des Kongresses Erasmus*, Amsterdam, London: North-Holland Publishing Company 1971a.

Schalk, Fritz: *Praejudicium im Romanischen.* Frankfurt a.M.: V. Klostermann 1971b.

Scheerer, Eckart: Sinne, die, in: Joachim Ritter (ed.): *Historisches Wörterbuch der Philosophie.* Basel: Schwabe 1995, vol. 9, 824–869.

Schleusener-Eichholz, Gudrun: Sehen, in: Ralf Konersmann (ed.): *Wörterbuch der philosophischen Metaphern*. Darmstadt: Wissenschaftliche Buchgesellschaft 2007, 368–375.

Schlobach, Jochen: *Zyklentheorie und Epochenmetaphorik*. München: Fink 1980.

Schmidt-Biggemann, Wilhelm: *Theodizee und Tatsachen*. Frankfurt a.M.: Suhrkamp 1988.

Schmitt, Arbogast: Die „Wende des Denkens auf sich selbst", in: Maria Moog-Grünewald (ed.): *Das Neue. Eine Denkfigur der Moderne*. Heidelberg: Winter 2002, 13–38.

Schrader, Ludwig: *Sinne und Sinnesverknüpfungen. Studien und Materialien zur Vorgeschichte der Synaesthesie und zur Bewertung der Sinne in der italienischen, spanischen und französischen Literatur*. Heidelberg: Winter 1969.

Schröder, Gerhart: *Baltasar Graciáns ‚Criticón'*. München: Fink 1966.

Schulte, Hansgerd: *Del desengaño. Wort und Thema in der spanischen Literatur des Goldenen Zeitalters*. München: Wilhelm Fink Verlag 1969.

Screech, Michael Andrew: An aspect of Montaigne's aesthetics. «Entre les livres simplement plaisans ... » (III, 10, Des livres), in: *BHR* 24, 1962, 576–582.

Seelmann, Kurt: *Theologie und Jurisprudenz an der Schwelle zur Moderne. Die Geburt des neuzeitlichen Naturrechts in der iberischen Spätscholastik*. Baden-Baden: Nomos 1997.

Séginger, Gisèle: Infusoires, monères et cellules: la science à l'épreuve de la littérature. De Michelet à Flaubert, in: Miroslaw Loba/Barbara Luczak (ed.): *Formes du vivant, formes de littérature*. Poznan: Naukowe 2019, 31–48.

Séginger, Gisèle: Transmission ou trahison? La circulation triangulaire des savoir du vivant (Michelet, Flaubert, Zola), in: Thomas Klinkert/Gisèle Séginger (ed.): *Littérature française et savoirs biologiques au XIXe siècle*. Berlin: De Gruyter 2020, 84–94.

Seibold, Eugen/Seibold, Ilse: Buffon (1707–1788) als geologischer Pionier, in: *International Journal of Earth Sciences* 98, 8, December 2009, 2023–2029.

Sihvola, Ari: *Ubi materia, ibi geometria*. Proceedings of Bianisotropics 2000. Lissabon, Portugal, 27.– 29. September 2000, 187–192.

Simson, Uwe: Spengler?, in: *Merkur* Heft 700, 61, 2007, 730–741.

Spaemann, Robert/Löw, Reinhard: *Die Frage Wozu? Geschichte und Wiederentdeckung des teleologischen Denkens*. München, Zürich: Piper 1986.

Stannard, Michael W.: *Galdós and Medicine*. Bern: Peter Lang 2015.

Steinbüchel, Theodor: *Der Zweckgedanke in der Philosophie des Thomas von Aquino*. Münster: Aschendorff 1912.

Stenger, Gerhardt: Quand Voltaire expliquait l'attraction Newtonienne aux français (à propos de la XVe Lettre philosophique), in: *Revue Voltaire* 13, 2013, 1–15.

Stenger, Gerhardt: Sur un problème mathématique dans la XVIIe Lettre philosophique, in: *Cahiers Voltaire* 5, 2006, 11–22.

Stockinger, Hermann E.: *Die hermetisch-esoterische Tradition. Philosophische Texte und Studien*. Hildesheim, Zürich: Georg Olms Verlag 2004.

Strauss, Leo: *Naturrecht und Geschichte*. Stuttgart: Koehler 1956.

Strolle, Jon M.: Engaño and Art in the ‚Criticón', in: *Hispanofila* 58, 1976, 5–17.

Strosetzki, Christoph: Der Luxus, das Notwendige und der Naturzustand. Kontroversen und Kontexte im Frankreich der Frühen Neuzeit, in: Annika Nickenig/Urs Urban (ed.): *Dinge – Gaben – Waren*. Berlin: Metzler, 2022a, 175–186.

Strosetzki, Christoph: Der Ratsherr im *Siglo de Oro*: ein Berufsbild zwischen Fürstenspiegel, bürgerlicher Emanzipation und literarischer Satire, in: D. Briesemeister/S. Große/A. Schönberger (ed.): *Dulce est philologiam colere. Festschrift*. Berlin: 1999, 657–677.

Strosetzki, Christoph: Fortschritt und Erfindung im Spannungsfeld von dignitas und miseria hominis, in: Die Idee von Fortschritt und Zerfall im Europa der frühen Neuzeit, in: *Germanisch-Romanische Monatsschrift*, Neue Folge 58, 1, 2008, 113–123.

Strosetzki, Christoph: *Handarbeit und Kopfarbeit. Humanistenwissen im Siglo de Oro.* Berlin: Metzler 2022b.

Strosetzki, Christoph (ed.): *Juan Luis Vives. Sein Werk und seine Bedeutung für Spanien und Deutschland.* Frankfurt: Vervuert Verlagsgesellschaft 1995.

Strosetzki, Christoph: *Literatur als Beruf. Zum Selbstverständnis gelehrter und schriftstellerischer Existenz im spanischen Siglo de Oro.* Düsseldorf: Droste 1987.

Strosetzki, Christoph: Philologie und Physik in der Romantik und heute, in: Christoph Strosetzki (ed.): *200 Jahre Nationalphilologie.* Berlin: Metzler 2022c, 271–279.

Strosetzki, Christoph: Verstehen bei Montaigne, in: *Wolfenbütteler Renaissance Mitteilungen* 6, 1, Hamburg 1982, 89–104.

Thiher, Allen: *Fiction Rivals Science: The French Novel from Balzac to Proust.* University of Missouri Press 2001.

Tränkle, Hermann: Gnothi seauton. Zu Ursprung und Deutungsgeschichte des delphischen Spruchs, in: *Würzburger Jahrbücher für die Altertumswissenschaft*,1985, vol. 11, 19–31.

Vatin, François/Auguste Comte: Les sciences d'application et la formation du peuple, in: *Revue philosophique* 4, 2007, 421–435.

Ventarola, Barbara: Der Experimentalroman zwischen Wissenschaft und Romanexperiment. Überlegungen zu einer Neubewertung des Naturalismus Zolas, in: *Poetica* 42, N. 3/4 2010, 277–324.

Videla, Gloria: *El ultraísmo. Estudios sobre movimientos poéticos de vanguardia en España.* Madrid: Gredos 1971.

Weber, Wolfgang E. J.: *Geschichte der europäischen Universität.* Stuttgart: Kohlhammer 2002.

Wedel, Christine von: *Das Nichtwissen bei Erasmus von Rotterdam.* Basel, Frankfurt am Main: Halbing & Lichtenhahn 1982.

Weiser, Jutta: *Poetik des Pathologischen. Medizin und Romanliteratur in Spanien (1880–1905).* Freiburg: Rombach 2013.

Welzel, Hans: *Naturrecht und materiale Gerechtigkeit.* Göttingen: Vandenhoeck & Ruprecht 1962.

Westerwelle, Karin: *Montaigne. Die Imagination und die Kunst des Essays.* München: Fink 2002.

Widmer, Paul: Niedergang, Untergang, in: Joachim Ritter/Karlfried Gründer (ed.): *Historisches Wörterbuch der Philosophie.* Darmstadt: 1976, vol. 6, 838–846.

Wigger, Anne: *Vom ‚matasanos‘ zum ‚médico perfecto‘. Zum literarischen Bild des Arztes im Spanien des 16. Jahrhunderts.* Berlin: Walter Frey 2001.

Wild, Markus: Montaigne als pyrrhonischer Skeptiker, in: Carlos Spoerhase/Dirk Werle/Markus Wild (ed.): *Unsicheres Wissen, Skeptizismus und Wahrscheinlichkeit 1550–1850.* Berlin: De Gruyter 2009, 109–133.

Wilson, Catherine: Sehen, in: Joachim Ritter (ed.): *Historisches Wörterbuch der Philosophie*. Basel: Schwabe 1995, vol. 9., 149–157.

Wolfers, Benedikt: *Geschwätzige Philosophie. Thomas Hobbes' Kritik an Aristoteles.* Würzburg: Königshausen & Neumann 1991.

Zöckler, Otto: *Die Lehre vom Urstand des Menschen, geschichtlich und dogmatisch-apologetisch.* Gütersloh: Bertelsmann 1879.

Zurbuchen, Simone: Zur Wirkungsgeschichte der beiden Diskurse, in: Johannes Rohbeck/Lieselotte Steinbrügge (ed.): *Jean-Jacques Rousseau: Die beiden Diskurse zur Zivilisationskritik*. Berlin: De Gruyter 2015, 83–103.

The manufacturer's authorised representative in the EU is Springer Nature Customer Service Centre GmbH, Europaplatz 3, 69115 Heidelberg, Germany. If you have any concerns regarding our products, please contact ProductSafety@springernature.com

Printed and bound by CPI Group (UK) Ltd, Croydon, CR0 4YY

24/04/2026

02096375-0006